III International Symposium on Fundamental Aspects of Rare-Earth Elements Exploration, Mining and Separation and Modern Materials Engineering (REE-2016)

MATEC Web of Conferences Volume 96 (2017)

Almaty, Kazakstan
15 - 18 August 2016

Editors:

Mishik Kazaryan
Anna Godymchuk
Liudmyla Rieznichenko

ISBN: 978-1-5108-3564-1

Printed from e-media with permission by:

Curran Associates, Inc.
57 Morehouse Lane
Red Hook, NY 12571

proceedings.com

Some format issues inherent in the e-media version may also appear in this print version.

Printed by Curran Associates, Inc. (2017)

For additional information, please contact EDP Sciences – Web of Conferences at the address below.

EDP Sciences – Web of Conferences
17, Avenue du Hoggar
Parc d'Activité de Courtabœuf
BP 112
F-91944 Les Ulis Cedex A
France

Phone: +33 (0) 1 69 18 75 75
Fax: +33 (0) 1 69 28 84 91

contact-edps@webofconferences.org

Additional copies of this publication are available from:

Curran Associates, Inc.
57 Morehouse Lane
Red Hook, NY 12571 USA
Phone: 845-758-0400
Fax: 845-758-2633
Email: curran@proceedings.com
Web: www.proceedings.com

TABLE OF CONTENTS

ISSN (electronic edition): 2261-236X

National Research Tomsk State University
Joint Stock Company "Volkovgeology"
http://www.ree-symposium.org/

III INTERNATIONAL SYMPOSIUM

"Fundamental aspects of rare-earth elements exploration, mining and separation and modern materials engineering" (REE-2016)

Almaty, Republic of Kazakhstan
15-18 August 2016

REE-2016 provides a high quality of professional presentations to scientists and engineers who wish to promote and communicate the interaction between technical advances in the field of rare and rare earth elements mining, modification, separation, research, and application.

The goal is to bring researchers and practitioners together to share the latest knowledge on rare and rare earth elements technologies.

Major Topics

- Problems of exploration, production and extraction of rare and rare-earth elements
- Methods and approaches to the separation and isolation of highly pure and rare earth elements
- Industrial technology of production and separation of rare and rare earth elements
- Economic aspects of technologies
- Materials based on rare and rare earth elements (application in metallurgy, catalysis, medicine, optoelectronics, etc.)
- Trace elements based on rare and rare earth metals in agricultural biotechnology

Round-table meetings will be held during Symposium

REE-2016 language: English, Russian

Venue

ALTYN KARGALY HOTEL
www.altyn-kargaly.kz

Organizing Committee Contact

Dr. Natalia Kosova
tel./fax: +7 3822 412319, cell : +7 923 424 4253
info@ree-symposium.org

Scientific & Organizing Committee

Editor Team

Prof. Mishik Kazaryan
P.N. Lebedev Physical Institute of the Russian Academy of Sciences, National Academy of Science, Yerevan, Armenia
E-mail : kazar@sci.lebedev.ru

Dr. Anna Godymchuk
Tomsk Polytechnic University, Tomsk, Russia, NUST "MISIS", Moscow, Russia, Derzhavin Tambov State University, Tambov, Russia
E-mail: godymchuk@mail.ru

Dr. Liudmyla Rieznichenko
F.D. Ovcharenko Institute of Biocolloidal Chemistry National Academy of Science, Kiev, Ukraine
E-mail: reznichenko_ls@mail.ru

Detailed information on http://www.ree-symposium.org

edp sciences

MATEC Web of Conferences

Statement of Peer review

In submitting conference proceedings to MATEC Web of Conferences, I certify to the Publisher that I adhere to the **Policy on Publishing Integrity** of the journal in order to safeguard good scientific practice in publishing.

1. All articles have been subjected to peer review administered by the proceedings editors.
2. Reviews have been conducted by expert referees, who have been requested to provide unbiased and constructive comments aimed, whenever possible, at improving the work.
3. Proceedings editors have taken all reasonable steps to ensure the quality of the materials they publish and their decision to accept or reject a paper for publication has been based only on the merits of the work and the relevance to the journal.

Title, date and place of the conference

III International Symposium "Fundamental Aspects of Rare-earth Elements Exploration, Mining and Separation and Modern Materials Engineering", 15-18 August 2016, Almaty, Republic of Kazakstan

Proceedings editor(s):

Prof. Mishik Kazaryan
Dr. Anna Godymchuk
Dr. Liudmyla Rieznichenko

Date and editor's signature 20 December 2016

Prof. Mishik Kazaryan Dr. Anna Godymchuk Dr. Liudmyla Rieznichenko

DOI: 10.1051/matecconf/20179600001

Extractive Separation of Scandium from Rare Earth Elements

Elena Zlobina *, *Akmaral* Ismailova, and *Khajdar* Tassibekov

Al-Farabi Kazakhstan National University, Almaty, Kazakhstan

Abstract. The extractive separation of scandium and rare earth metals (REM) was studied. The extractants used are trialkylbenzylammonium di-2-ethylhexylphosphate (TABAC-D2EHPA, binary extractant) and trioctylphosphine oxide (TOPO). The effect of concentration and nature of mineral acids on the scandium recovery was studied. Changing the HNO_3 concentration in the range of 0.01 – 2.0 mol/L does not affect the extraction of metals by binary extractant, extraction corresponds to the following row: La, Eu, Gd < Tb <Y < Er < Yb < Sc, Th, that allows to separate scandium from rare earth metals by single extraction. Extraction of scandium by trioctylphosphineoxide from HCl solution proceeds quantitatively in the range of 0.5 – 7.0 mol/L, an increased HNO_3 concentration leads to the decrease in scandium extraction. The conditions for the extractive separation of scandium and REM using TOPO solution were determined: scandium can be isolated selectively at the HNO_3 concentration of 0.5 mol/L REM are transferred into the organic phase at the HNO_3 concentration of 5.0 mol/L. It was determined that the organic phase extracts the compounds of the following composition: (ScA_3 + $R_3R'NCl$) (HA - D_2EHPA, $R_3R'NCl$ - TABAC) and [$Sc(NO_3)_3 \cdot 3TOPO$].

1 Introduction

The value of scandium and its compounds in the creation of new high-tech materials is continuously increasing. Scandium is used in the production of high-temperature ceramics, superhard materials, materials for a new generation of computers, laser crystals, high quality optical glass, alloys for the aerospace industry, and others [1, 2]. The Republic of Kazakhstan has significant reserves of rare earth metals (REM). Industrially important content of scandium, yttrium, and rare-earth metals has been observed in uranium, titanium, zirconium, coal, phosphorus, aluminium deposits [3].

A variety of scandium-containing raw materials along with low target component, the complexity of scandium separation from the accompanying elements, especially close to REM in properties, does not allow the use of a universal method to extract it. After the acid or alkaline methods of the materials preparation in the course of the process chain there is a gradual concentration of scandium and rare-earth metals, which are mainly due to the use of extraction and precipitation processes. The process results in the concentrate, which is then separated into individual fractions or individual substances [4]. For scandium there is no selective extracting agent, which allows separating it along a large group of rare earth metals. There is only the way to create conditions under which scandium can be recovered with one or more cations.

The same neutral and cation-exchange reagents are used as industrial extractants for scandium and rare earth elements [5]. The degree of scandium extraction by di-(2-ethylhexyl)phosphoric acid (D2EHPA) is more than 90 % of the solution with HCl concentration of 1 mol/L, but most of yttrium and the lanthanides remain in the raffinate [6]. The main disadvantage of D2EHPA is non-selectivity: V, Fe, Zn, Cu, Co, Ni, Mn, Mg, Ca, together with scandium are transferred into the organic phase, moreover, solutions with high acid content must be used for back-extraction. Commercial extractants are used for separation of scandium from REE in the industry: di(2,4,4-trimethylpentyl)phosphinic acid (Cyanex272 bis(2,4,4-trimethylpentyl)phosphinic acid) and Cyanex302 (bis(2,4,4-trimethylpentyl) monothiophosphinic acid), which is monothioanalogue of Cyanex272 [7].

Monocarboxylic acids also possess low selectivity; they are used for group concentration operations. Increase the extraction selectivity can be due to the use of mixtures of reagents, careful selection of extraction conditions (pH, temperature, salting-out agents, thinners), as well as carrying out multi-stage processes. Authors of [8] used the solution of naphthenic acids and iso-octanol in kerosene as an extractant for separating scandium from REM. Twice extraction cycle allowed to achieve the separation factors for scandium from REE above 10^4. During the first cycle, which includes 10 stages, scandium was concentrated from 0.02 – 0.04 % to 15 – 20 % of Sc_2O_3. In the second cycle, which includes 3 stages of extraction and back-extraction, the resulting scandium concentrate was purified to 99.99 – 99.999 %.

The results of extractive separation of REM, yttrium and scandium using alkylphosphine oxides of different composition from Cyanex 921, Cyanex 923, and Cyanex 925 trademarks as extractants from

* Corresponding author: zeva65@mail.ru

solutions of nitric, hydrochloric, sulfuric and phosphoric acids are given in [9-11]. It was shown that REM can be divided into four groups. The compositions of extracted complexes were determined, extraction mechanism was proposed, and extraction constants were calculated. It was shown that REM extraction efficiency increases in the following order: Cyanex 921 > Cyanex 923 >> Cyanex 925.

The use of binary extractants (BE), which are ion pairs formed by an organic base and an organic acid, is a promising direction in the extraction technology of some metals. The main advantages of binary extractants are: possibility of the prediction of BE properties based on the data on initial extraction systems; simplicity of obtaining the ion pair and availability of the extractants; increase in the distribution and separation coefficients; simplification of the back-extraction; high rates of extraction and back-extraction [12]. Dependences of binary extraction of REM were investigated by the authors of [13]. It is shown that the extractant selectivity for the cation depends on the nature of the organic acid, containing in the extractant. Adjacent pairs of lanthanides can be separated with a separation factor of 2.5. It should be noted that some authors [14] called alkylammonium salts of various organic acids as ionic liquids or functionalized ionic liquids.

The problem of separation of scandium and rare earth metals is still not solved. The same extraction reagents are used for the extraction and separation of these metals. Since the new industrial reagents are not often created, the main objectives of a qualitative change in the technological schemes for REM and scandium are more complete study of existing reagents, selection of separation conditions, use of mixtures of extractants and their compounds, study of the complex-forming properties of reagents.

The goal of this work was to study the extraction patterns and possible separation of scandium and rare earth metals by binary extractant - trialkylbenzylammonium di-2-ethylhexylphosphate and solution of trioctylphosphine oxide in kerosene.

2 Experimental part

2.1 Reagents, equipment, methods of analysis

Standard solutions of scandium, yttrium and rare earth metals prepared according to standard procedures by dissolving the metal oxides in the appropriate acids were used. Working solutions of metals were prepared by diluting standard solutions to the desired concentration.

Solutions obtained after the microwave decomposition of mineral raw materials containing scandium, yttrium and rare earth elements were also used.

The content of all metals in the aqueous phase before and after extraction was determined by mass spectrometry with inductively coupled plasma (ICP-MS) using Agilent 7500a mass spectrometer. Operating parameters of the spectrometer are as follows: plasma gas flow rate - 15 L/min; carrier gas flow rate – 0.8 - 1.3 L/min; plasma temperature - 8000 °C; signal integration time - 0.1 seconds; high-frequency signal power - 1200-1400 W. Calibration solutions for mass spectrometric analysis were prepared from Multi-Element Calibration Standard-1 and Multi-Element Calibration Standard-2A standard solutions (Agilent Technologies). Nitric acid of high purity grade was used for preparation of the solutions; all solutions were prepared using twice-distilled water.

The pH was adjusted using sodium hydroxide, ammonia and hydrochloric acid solutions. Hydrogen ions concentration was monitored using I-160MI ionomer with ES-10603 glass electrode standardized by buffer solutions.

2.2 Extractants, extraction technique

Trioctylphosphine oxide $(C_8H_{17})_3PO$ (TOPO) and binary extractants were used as extraction reagents.

For preparation of binary extractants D2EHPA, di-2-ethylhexyldithiophosphoric acid (D2EHDTPA), trioctylamine (TOA) and trialkylbenzylammonium chloride (TABAC) were used. Ion pairs obtained by contacting solutions of TABAC ($R_3R'NCl$) with organic acids in heptane in 1:1 ratio with 2 mol/L sodium hydroxide solution for 10 minutes. Contacting the organic phase and the alkaline solution was repeated three times, refreshing NaOH solution. After separation, the organic phase was washed with water to neutral pH. Ion pairs TOA – D2EHPA obtained by contacting solutions of TOA ($(C_8P_{17})_3N$) with D2EHPA in heptane in 1:1 ratio for 10 minutes. Composition of the BE obtained is given in Table 1. Kerosene and heptane were used as diluents.

The technique for metals extraction was as follows: the exact volume of metal solution was placed into a separatory funnel having a certain value of the acid concentration, an aliquot of an extractant was added, and the extraction was carried out by shaking the separatory funnel by hand for 5 minutes. Preliminary experiments showed that this time is enough to establish equilibrium in the system. All experiments were performed at 25 ± 1 °C. Further, the phases were separated; the metal content in the water phase was controlled using ICP-MS method. The metal content in the organic phase was determined by the difference between its content in the solution before and after extraction.

Distribution coefficients (D) were calculated according to the equation:

$$D = \frac{C_o}{C_a}, \quad (1)$$

where C_o and C_a are metal concentrations in the organic and aqueous phases.

Recovery rate (R,%) was calculated based on the relationship:

 DOI: 10.1051/matecconf/20179600001

$$R = \frac{D}{D + V_a/V_o} \cdot 100\ \%, \qquad (2)$$

where V_a and V_o are the volumes of the aqueous and organic phases.

Metal separation efficiency was evaluated by calculating the value of the enrichment degree (r), showing each metal fraction in the total amount of metals in the solution. The enrichment degree was calculated using the formula:

$$r = \frac{C_{Me}}{C_{total}} \cdot 100\ \%, \qquad (3)$$

where C_{Me} is metal concentration, mg/dm^3, C_{total} is the sum of the concentrations of all metals in the solution, mg/dm^3.

Table 1. Composition of binary extractants.

Designation	Initial reactants	Composition of binary extractant
BE-1	TABAC – D2EHPA	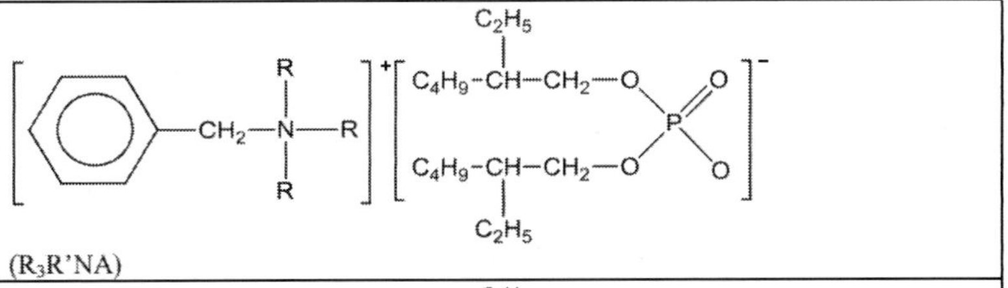 (R$_3$R'NA)
BE-2	TABAC – D2EHDTPA	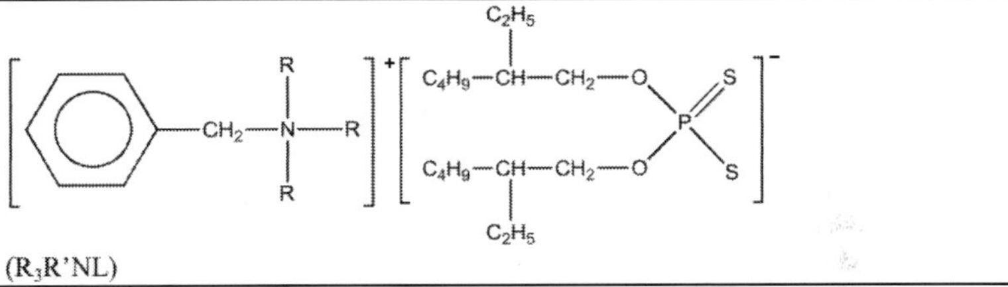 (R$_3$R'NL)
BE-3	TOA – D2EHPA	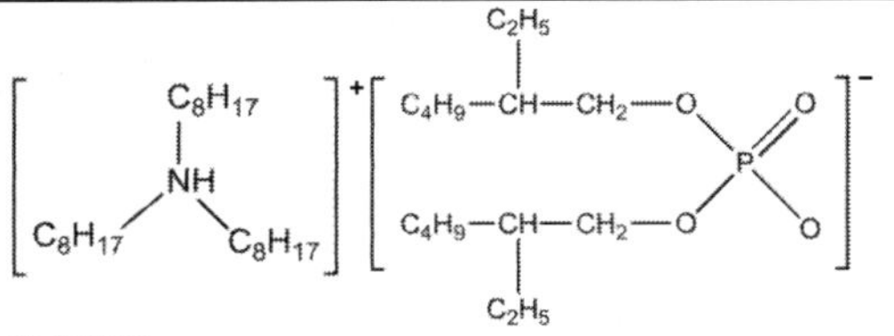 (R$_3$NHA)

3 Results and discussion

3.1 Extraction of scandium using binary extractants of different composition

Extraction behavior of binary extractants can be predicted when considering the patterns of the metals extraction by reagents used for the preparation of BE. In the nitrate solutions scandium is in the form of a hydrated $[Sc(H_2O)_6]^{3+}$ cation, therefore it can be extracted effectively by cation exchange extractants. BE properties are related to the properties of the initial systems, so cation exchange extractant, which is part of the binary extractant affects the degree of extraction of scandium. Previously, it was found that D2EHPA effectively (over 95 %) extracts scandium in a wide range of the aqueous phase acidity; the degree of extraction using D2EHDTPA does not exceed 80 – 95 % [15].

Extraction of scandium using BE of different composition was studied depending on the concentration of HNO_3, which was varied in the range of 0.01 – 3.0 mol/L (Fig. 1).

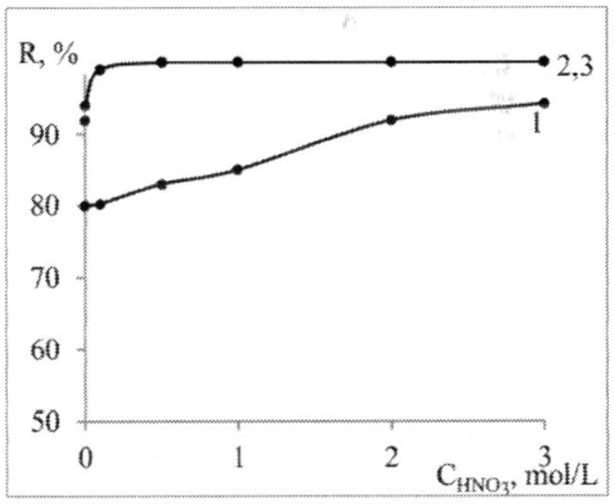

Fig. 1. Extraction of scandium using BE of different composition: 1 – TABAC-D2EHDTPA; 2 – TOA-D2EHPA; 3 – TABAC-D2EHPA; $C_{Sc} = 1.0 \cdot 10^{-4}$ mol/L; C_{ex} = 0.1 mol/L

From these data, it follows that BE containing TABAC-D2EHPA and TOA-D2EHPA quantitatively extract scandium from solution with the nitric acid concentration of 0.1 – 3.0 mol/L. Extracting scandium using BE containing TABAC-D2EHDTPA has been increased from 80 to 95 % at HNO_3

concentration increase from 0.01 to 3.0 mol/L. Binary extractant of TABAC-D2EHPA composition was selected for further research.

3.2 Study of the effect of mineral acid concentration

The main method of decomposition of scandium-containing natural and man-made samples is the dissolution in mineral acids, therefore, the effect of HCl and HNO_3 concentrations on the extraction of scandium by binary extractants and TOPO was studied. The results are shown in Figures 2 and 3.

The degree of cation extraction in binary extraction of metal salts is affected by the nature of the anion. Obtained data shows that the extraction of scandium by BE of TABAC-D2EHPA composition from nitric acid solutions proceeds more effectively than from solutions of hydrochloric acid in the concentration range of 0.1 – 3.0 mol/L.

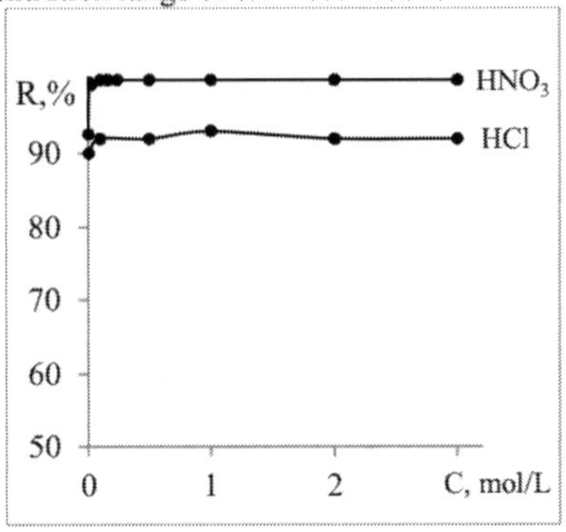

Fig. 2. Extraction of scandium by 0.1 mol/L solution of TABAC-D2EHPA in heptane depending on the concentration of mineral acid: $C_{Sc} = 1.0 \cdot 10^{-4}$ mol/L; O:W = 1:10; τ = 5 min; T = 25 °C.

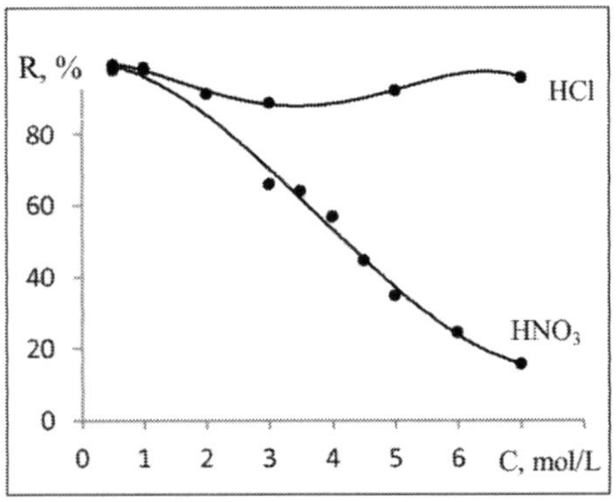

Fig. 3. Extraction of scandium by 0.01 M solution of TOPO in kerosene depending on the concentration of mineral acid: $C_{Sc} = 1.0 \cdot 10^{-4}$ mol/L; O:W = 1:10; τ = 5 min; T = 25 °C.

The findings are consistent with the extraction raw for the initial anion exchange systems with QAB salts. Binary extraction of scandium is not dependent on the solution pH, unlike cation exchange systems, since the increase in the concentration of hydrochloric acid remains the extraction degree of scandium unchanged.

Extraction of scandium using TOPO-kerosene solution from hydrochloric acid solutions proceeds more efficiently than from nitric acid solutions in the concentrations range of 2.0 – 7.0 mol/L. Increasing HNO_3 concentration reduces the extraction of scandium.

3.3 Study of the effect of scandium concentration in the aqueous phase

Metal ion species in the aqueous and organic phases are different, so the metal distribution ratios depend on the initial concentration of the element in the aqueous phase. Scandium extraction by TABAC-D2EHPA-heptane and TOPO-kerosene solutions was performed varying the concentration of metal in the aqueous phase. Figures 4 and 5 show the isotherms of scandium extraction.

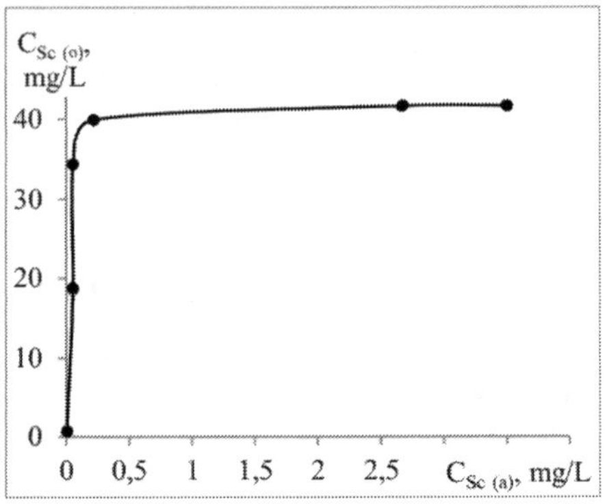

Fig. 4. The isotherms of scandium extraction by binary extractant TABAC-D2EHPA-heptane: C_{BE} = 0.01 mol/L; C_{HNO_3} = 1.0 mol/L; O:W= 1:1; τ=10 min.

The capacity of 0.01 M TABAC-D2EHPA for scandium is 40 mg/L. Distribution of scandium in binary systems (excluding the possible interactions in the organic phase) can be described by the following scheme:

$$3R_3R'NA_{(o)} + Sc^{3+}_{(a)} + 3NO_3^-{}_{(a)} \rightarrow$$
$$\rightarrow ScA_{3(o)} + 3[R_3R'N]NO_{3(o)}$$

Capacity of 0.001 M solution of TOPO in kerosene for scandium is ~7.7 mg/L. As can be seen from the figure, when the scandium concentration in the aqueous phase is more than 12 mg/L isotherm line becomes parallel to the abscissa, indicating the scandium saturation of extractant.

The process of scandium extraction by TOPO-kerosene solution can be represented by the scheme:

$$Sc^{3+}{}_a + 3NO_3^-{}_a + 3TOPO_o = [Sc(NO_3)_3 \cdot 3TOPO]_o$$

Equation for scandium extraction constant by TOPO-kerosene solution has the following form:

$$K_{ex} = \frac{[Sc(NO_3)_3*3TOPO]_o}{[Sc^{3+}]_a[NO_3^-]_a^3[TOPO]_o^3} \qquad (4)$$

$$\lg K_{ex} = 5{,}84$$

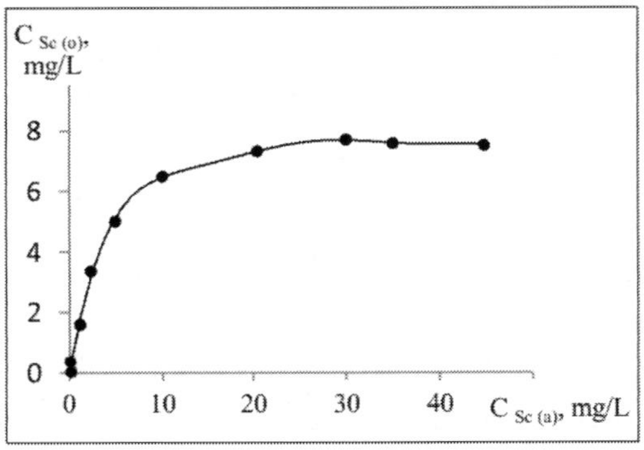

Fig. 5. The isotherms of scandium extraction by TOPO-kerosene: C_{TOPO} = 0.001 mol/L; C_{HNO_3} = 0.5 mol/L; O:W = 1:1.

3.4. Study of the separation of scandium and rare-earth metals

Extractive separation of scandium and rare-earth elements was carried out using 0.1 mol/L solution of TABAC-D2EHPA binary extractant in heptane, the concentration of metals was 400 μg/L.

The concentration of nitric acid in the aqueous phase was varied in the range of 0.01 – 2.0 mol/L. Dependence of the extraction degree from the nitric acid concentration is shown in Figure 6.

Varying the concentration of nitric acid in the range of 0.01 – 2.0 mol/L has almost no effect on the extraction of scandium, thorium, yttrium, rare-earth metals at the joint presence. Scandium and thorium was transferred quantitatively into the organic phase (> 95%).

The degree of Yb extraction does not exceed 58 %, Er, Y – 30 %, Tb, La, Gd, Eu – 18 %. Extraction corresponds to the following row: La, Eu, Gd < Tb < Y < Er < Yb < Sc, Th, which is consistent with the cation exchange row of D2EHPA, but in the case of binary extraction recovery rates are constant with changes in nitric acid concentration. This allows to obtain concentrates of scandium, yttrium and rare-earth metals with the ratio specified in Table 2 at the nitric acid concentration of 0.01 - 2.0 mol/L.

Table 2. Degree of enrichment by rare-earth metals during the extraction by binary extractant of TABAC-D2EHPA composition.

	Degree of enrichment, %								
	Sc	Y	La	Eu	Gd	Tb	Er	Yb	Th
Without acid	11.0	11.3	11.3	11.2	11.2	11.3	11.5	11.1	10.2
In HNO_3 solution	32.3	7.4	1.0	1.2	1.4	2.2	9.5	15.6	29.3

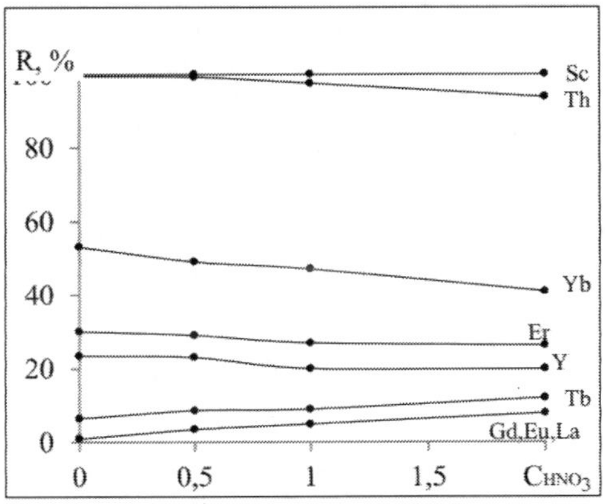

Fig. 6. Extraction of metals by TABAC-D2EHPA at the joint presence: C_{ex} = 0.1 mol/L; C_{Me} = 400 μg/L.

The recovery of scandium by TOPO-kerosene solution at $C(HNO_3)$ = 7.0 mol/L does not exceed 15 %, which suggests that at high HNO_3 concentrations scandium and related metals can be separated. Therefore, extractive separation of metals was studied using model solution by TOPO-kerosene extractant varying nitric acid concentrations in the range of 0.5 - 5.0 mol/L.

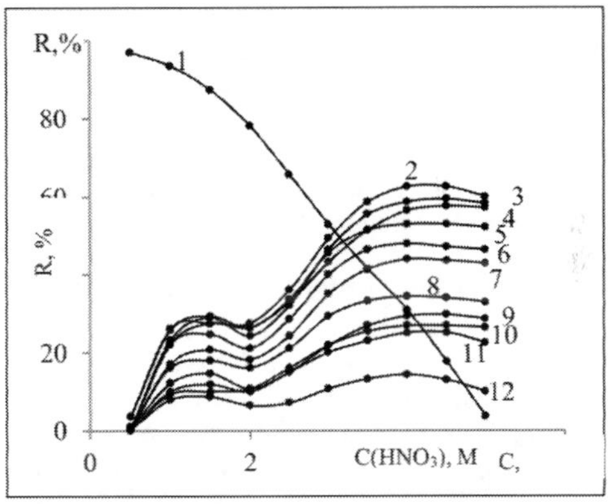

Fig. 7. Extraction of scandium and REM by 0.01 M solution of TOPO-kerosene depending on the concentration of HNO_3: 1 – Sc; 2 – Yb; 3 – Er; 4 – Dy; 5 – Sm; 6 – Gd; 7 – Nd; 8 – Eu; 9 – Lu; 10 – Tm; 11 – Ho, Tb, Pr, Ce, La; 12 – Y; C_{Me} = 40 μg/L, C_{Sc} = 450 μg/L.

Besides scandium model solution contained the following rare-earth metals: Ce, Pr, Nd, Sm, Eu, Gd, Tb, Dy, Ho, Er, Tm, Yb, Lu, Th, Y and La, moreover REM concentration was 40 μg/L, concentration of Sc was 450 μg/L. The results are shown in Figure 7.

As can be seen from the data, the degree of scandium extraction is reduced with increasing concentration of nitric acid, and not exceeding 5 % at $C(HNO_3)$ = 5.0 mol/L, at the same time the extraction of rare-earth metals increases. Thus, when nitric acid concentration is 0.5 mol/L, it is possible to separate scandium from rare earth impurities using TOPO-kerosene solution; rare earth impurities remain in the aqueous phase under given conditions. However, if the nitric acid concentration is 5.0 mol/L rare-earth metals impurities are transferred into the organic phase unlike scandium. REE can be extracted quantitatively under these conditions by performing multiple extractions by TOPO-kerosene solution. In this case, it is possible to manage the production stage without back-extraction of scandium from the organic to the aqueous phase. This will reduce the number of operations and the amount of reagents used.

4 Conclusion

The extraction patterns of scandium by binary extractants of different composition have been studied: trialkylbenzylammonium di-2-ethylhexyldithiophosphate, trialkylbenzylammonium di-2-ethylhexylphosphate, trioctylammonium di-2-ethylhexylphosphate and solution of trioctylphosphine oxide in kerosene.

The effect of concentration and nature of mineral acids on the extraction of scandium have been studied. Binary extractants based on D2EHPA allow extracting scandium from the solution most effectively, the presence of TABAC branched cation makes the binary extractant selective. The extraction of scandium by binary D2EHPA-TABAC extractant depends on the nature of the mineral acid in aqueous solution. From the solutions of nitric acid with concentrations of 0.1 – 3.0 mol/L scandium is extracted quantitatively, the recovery from solution of hydrochloric acid does not exceed 90 %. The nature of mineral acid affects the scandium extraction by TOPO-kerosene solution. In the wide range of HCl concentration, 0.5 –7 mol/L, scandium is extracted quantitatively into the organic phase. With increase in HNO_3 concentration scandium extraction is reduced not exceeding 15 % at $C(HNO_3)$= 7 mol/L.

Based on the extraction isotherms the capacities of the extractants by scandium were determined, which are 40 mg/L for 0.01 mol/L TABAC-D2EHPA and 7.7 mg/L for 0.001 mol/L TOPO solution. It was determined that the organic phase extracts the compounds of the following composition: (ScA_3 + $R_3R'NCl$) and $[Sc(NO_3)_3 \cdot 3TOPO]$.

Study of scandium and REM separation showed that during the binary extraction scandium and thorium are transferred into the organic phase quantitatively (over 95 %) and the extraction degree of Yb does not exceed 58 %, Er, Y – 30 %, Tb, La, Gd, Eu – 18 %. Extraction corresponds to the following row: La, Eu, Gd $<$ Tb $<$ Y $<$ Er $<$ Yb $<$ Sc, Th, which is consistent with the cation exchange row of D2EHPA, but in the case of binary extraction recovery rates are constant with changes in nitric acid concentration. The conditions for the extractive separation of scandium and REM using TOPO solution were determined. Scandium can be isolated selectively into the organic phase at the HNO_3 concentration of 0.5 mol/L, REM are transferred into the organic phase at the HNO_3 concentration of 5.0 mol/L.

References

1. A.V. Naymov, Proceedings of Higher Schools Nonferrous Metallurgy (in Russian – Izvestiya Vuzov. Tsvetnaya Metallurgiya) **1**, 22 (2008)
2. S. Jacenko, V. Diev, B. Ovsjanikov, Eurasian Metals (in Russian – Metally Evrazii) **4**, 60 (2004)
3. N.S. Bekturganov, M.A. Najmanbaev, T.Ju. Surkova, Non-ferrous Metals (in Russian – Tsvetnye Metally) **4**, 48 (2010)
4. W. Weiwei, P. Yoko, Y.C. Chu, Hydrometallurgy **108**, 100 (2011)
5. X. Feng, T.A. Zhang, D. Dreisinger, F. Doyle, Miner. Eng. **56**, 10 (2014)
6. M. Ochsenkuhn-Petropulu, Th. Lyberopulu, G. Parissakis, Anal. Chim. Acta. **315**, 231 (1995)
7. M. Karve, Bh. Vaidya, Sep. Sci. Technol. **43**, 1111 (2008)
8. C. Liao, G. Xu, J. Jia, Y. Zhang, S. Wu, C. Yan, J. Chin. Rare Earth Soc. **19**, 289 (2001)
9. B. Gupta, A. Malik, A. Deep, Solvent Extr. Ion Exch. **21**, 239 (2003)
10. D. Li, C. Wang, Hydrometallurgy **48**, 301 (1998)
11. Y.A. El-Nadi, N.E. El-Hefny, J. A. Daoud, Solvent Extr. Ion Exch. **25**, 225 (2007)
12. A.I. Hol'kin, Chemical Technology (in Russian – Khimicheskaya Tehnologiya) **5**, 3 (2000)
13. V.V. Belova, A.A. Voshkin, A.I. Kholkin, A.K. Payrtman, Hydrometallurgy **97**, 198 (2009)
14. X. Sun, H. Luo, S. Dai, Talanta, **90**, 132 (2012)
15. E.V. Zlobina, L.O. Pak, Chemical Bulletin of Kazakh National University (in Russian – Vestnik KazNU. Suriya Himicheskaya) **4**, 168 (2013)

Acid Activation of Natural Zeolite with High Content of Iron Oxides in Creation of Selective Sorbents and Catalysts

Kairat Kadirbekov[1,2,*], *Dauren* Zhambakin[1], *Almaz* Kadirbekov[1,2], and *Kylysh* Imanbekov[2]

[1] LLP "Kazatomprom-Sorbent" Almaty, Kazakhstan
[2] JSC "A.B. Bekturov Institute of Chemical Sciences", Almaty, Kazakhstan

Abstract. The paper studies the influence of the nature of modifying acids (mineral, organic and heteropolyacids), and their combination on the composition and structure of the iron oxide rich clinoptilolit from Shankanay field in Kazakhstan for creation of selective catalysts in hydracarbon processing and sorbents for extracting ions of lanthanide and actinide elements. It is shown that sequential processing of natural zeolite in optimal conditions, by hydrochloric and sulfosalicylic acids lead to intensive decationization and dealumination, as well as maximum removal of iron ions from the zeolite framework without destroying it. It is found that the combination of activated clinoptilolite with hydrochloric acid and phosphotungstic heteropolyacid contributes to obtain catalyst system with high surface area and acidity.

1 Introduction

Zeolites are versatile catalyst systems [1-3]. They are processed using the most diverse hydrocarbon feedstock and produced to a wide range of commercially important products – components of fuels and synthetic oils, monomers and intermediates of petrochemical synthesis and others. In addition, modern sorption technology based on zeolite obtained qualitatively new sorbents with increased capacity and high selectivity of action. In [4-12] shows the prospects of their application in many catalytic and sorption processes.

In the last 10 years, authors are performing intensive research on the construction of catalytic systems and the creation of sorbents for the extraction of metals by modifying the natural zeolite Shankanay field (Kazakhstan) [13-15].

Shankanay natural zeolites have a high content of iron oxides, along with oxides of alkali and alkaline earth metals in their composition. It is generally known that for the selective removal of ferric ions which catalyze the carburization process in the processing of hydrocarbons using zeolite chelating treatment is carried out with these organic acids. It is also known that the need to strengthen the forces of acid centers, which is cracking reaction, it is recommended to run the process with the use of heteropolyacids or superacids.

At the same time, the effect of sequential treatment of natural zeolite with mineral then organic acids or mineral acid then heteropolyacids on activity, selectivity and catalyst stability of studied systems in specific chemical reaction (cracking hydrocarbons) has been studied insufficiently. In addition, the development of new catalysts for petrochemical processes or special sorbents for the extraction of rare elements require a large amount of systematic research work aimed at the establishment of physical and chemical laws of formation of complex catalytic systems according to the properties of their individual components, phase composition, structure, and so on. The solution of these important issues is dedicated to the proposed work.

2 Experimental Part

As the object of research natural zeolites of Shankanay field (Kazakhstan) was taken, in which the major rock-forming mineral is acid- and heat-resistant clinoptilolite.

Sorbents and catalysts based on natural zeolite of Shankanay field were synthesized by modification with mineral acids, organic acids, and heteropolyacids (HPA).

Handing out of zeolite by mineral acids was performed in order to remove the cations of various metals from the surface (primarily alkali and alkaline earth). Thus, one-, two- and three-time modification of natural zeolite by 1.75N hydrochloric acid following cracking catalysts were prepared: HKl-1, HKl-2 and HKl-3.

Treatment of zeolite by organic acids was carried out for the selective removal of ferric ions which catalyze the process of carbonization. To remove Fe^{3+} ions chelating agents were used: citric acid and sulfosalicylic acid (10 and 20 % solutions), ethylenediaminetetraacetic acid and its disodium salt EDTA (Trilon B, 10% aqueous solution). All of them form with iron ions stable water-soluble complexonates. Treatment by chelating agents were performed on natural (Kl) and pre decationized by hydrochloric acid (HKl-1) samples.

Treatment by HPA was carried out for enhancing the force of acidic centers at which the cracking reaction occurs. For that reason, new cracking catalysts prepared

* Corresponding author: kkairati@mail.ru

by modifying the decationized natural zeolite with heteropolyacids of molybdenum and tungsten series: PMo_{12}-HPA/HKl and PW_{12}-HPA/HKl. HPA was applied on the surface of decationized natural zeolite (HKl) in an amount of from 1 to 10% by the well-known impregnation technique.

Modification of natural zeolite samples (Kl) was performed by the method of Kerr in a Soxhlet apparatus. When heated, the cartridge with an acid placed in an extractor, while aqueous suspension of the zeolite placed in a water bath heated flask. And the "direct" method performed by placing the cartridge with the acid directly into a flask with acidified zeolite suspension. All samples were dried in an oven and calcined at 500 ° C for 4 hours in a muffle furnace.

An acid modification was conducted for 1.6-2 mm fraction of the zeolite (the ratio S:L of 1:10 wt.) at the processing temperature 94-98°C during 10 hours. After a predetermined time, the samples were washed thoroughly from residual acid and dried to air-dry state.

Oxide and elemental composition of samples of Shankanay field was identified by emission spectral analysis on diffractometer DFS-13.

X-ray analysis of natural zeolite and its' modified forms was performed on a D8 Advance (Bruker), α-Cu, tube voltage 40 kV, current 40 mA. Data processing and calculation of the diffraction patterns, calculation of interplanar distances were performed using the software EVA. Deciphering of samples and phases search were carried out under the Program Search / Match with Database of powder diffractometric data PDF-2 Rel. 2012 (ICDD).

IR study of samples was conducted by the spectrometer NICCOLET-2700 in the frequency range 400 - 4400 cm^{-1}. The catalyst was formed into a tablet "thickness" of 60 - 100 mg/cm^2. Adsorption of test gases was carried out at different temperatures, vacuuming 10-5 Torr.

3 Results and Discussion

3.1 Modifying the natural zeolite with an inorganic acid

In order to produce modified forms of Shankanay zeolites applied acid activation technique was applied.

Acid resistance of natural zeolites is essential for their technological applications, as well as the study of conditions for the preservation of different samples crystallinity of acid activated clinoptilolite will create a basis for the preparation of intermediates in the degree of dealumination and decationization of samples that have acidic properties, which play a major role in the catalytic cracking, and sorbents with high capacity volume.

It was determined that by treating 1.5N; 1.75N (optimum concentration) and 2.0N hydrochloric acid for 6 hours at a temperature of 96-98°C clinoptilolite was subjected to dealumination and decationization without appreciable destruction of the crystal lattice. Thus, acid treatment results in changes in the chemical composition of samples of natural zeolite in the process of modification: with increasing concentration of the activating acid simultaneously reduced oxides content of divalent metals and iron oxide in HKl samples.

However, on the degree of dealumination and decationization, apart from acid concentration, affect temperature, the amount of the acid solution, the duration and frequency of treatment. Consequently, on the basis of natural zeolite of Shankanay field following modified forms were obtained: treating the clinoptilolite by 1.75 N hydrochloric acid (designated as sample HKl-1), twice processing fresh portions of the same concentration of acid (HKl-2) and three-time processing of fresh portions of 10% hydrochloric acid (HKl-3).

The composition of the prepared modified forms of natural zeolite samples, as determined by emission spectral analysis is shown in Table 1.

Table 1. The oxide composition of samples modified by mineral acids.

Oxide	Kl	HKl-1	HKl-2	HKl-3
Al_2O_3	8.5	5.7	4.5	2.4
SiO_2	65.0	80.0	78.0	89.0
Fe_2O_3	10.2	2.7	3.5	1.2
CaO	3.5	0.7	0.2	0.1
MgO	2.5	1.1	0.3	0.5
Na_2O	2.1	1.8	0.4	1.2
K_2O	2.3	1.8	0.1	1.6
SiO_2/Al_2O_3	13.1	23.8	29.5	63.0

When treating zeolite by hydrochloric acid, content of mono- and divalent metal oxide and iron oxide, carbon ions which catalyze carburization, reduces from 10.2 wt%. in natural to 2.7 wt%. in acid activated sample. The amount of iron oxide, as compared with the quantity of the sample HKl-1 (2.7%), doubly treated zeolite sample with 1.75 N hydrochloric acid is 3.5%, and three times treated with 10% hydrochloric acid solution is reduced to 1.2%. The amount of alumina varies from 8.5 to 2.4%, i.e. reduced by 72%, whereas the SiO_2 content in samples of zeolite increases from 65.0 to 89.0%. This leads to increase of the value of silica modulus (molar ratio of SiO_2/Al_2O_3), with the first 13.5 (HKl-1) to 23,75 (HKl-2), then rapidly to 63 (HKl-3). The observed increase of silica modulus from 13.5 to 23.75, by which indirectly judge the acidic properties of the samples indicates the occurrence of acid sites. Under the influence of acids with low concentration there is exchange of zeolite cations into protons to form hydrogen form of zeolite (Broensted acid site). With the increase of acid concentration is obtained aluminum form by switching from aluminum tetrahedral coordination in the exchange position of the three-coordinated (Lewis Center).

It should be noted that excessively removing aluminum oxide from clinoptilolit (more than 50%) reduces the strength of the crystal structure of clinoptilolite. Indeed, by increasing the multiplicity of hydrochloric acid treatment of the clinoptilolit, dealumination and the formation of the amorphous phase is observed.

As an example, in Figure 1, HKl-1 is sample of natural zeolite treated by 1.75 N hydrochloric acid is presented.

X-ray picture of the zeolite sample (Fig.1) shows that the main minerals which form a solid framework of the zeolite is clinoptilolite with formula $Na_8(Al_6Si_{30}O_{72})(H_2O)_9$. Usual crystal-clinoptilolite formula: $(Na,K)6[Al_6Si_{30}O_{72}]*24H_20$. Together with clinoptilolite accompany minerals: analcime $Na_{16.24}Al_{16.00}Si_{32.00}O_{96}$ $(H_2O)_{16}$ (chemical formula $Na[AlSi_2O_6]{\bullet}H_2O$), quartz ($SiO_2$) and feldspars. Feldspars are presented in the form of solid solutions of the ternary system isomorphous series $K[AlSi_3O_8]$ — $Na[AlSi_3O_8]$ — $Ca[Al_2Si_2O_8]$ (or K-Na-, Ca-Na-feldspars), in particular in the form of albite $Na[AlSi_3O_8]$. From the remaining minerals it can be identified widespread iron minerals - hematite Fe_2O_3 and siderite $FeCO_3$.

Comparison of X-ray data of initial and acid activated samples pointed on changes in the structure of the mineral acid in the process of acid activation.

It was revealed that after treatment by 1.75N acid, mineral structure remained unchanged, but compared with the diffraction pattern of the initial sample of natural zeolite, there was a decrease of the intensity of one of the main reflections characteristic of the mineral (8.92 Å), as well as an increase in the reflection characteristic amorphous phase (3,3 Å).

When HKl sample was treated with 1.75N HCl acid there was dissolution of quartz, feldspar, and their value reduced by 20-25 %. Furthermore, in the HKl sample, there is a decrease of phases with iron oxide minerals: hematite content decreases, siderite dissolved by the mineral acid.

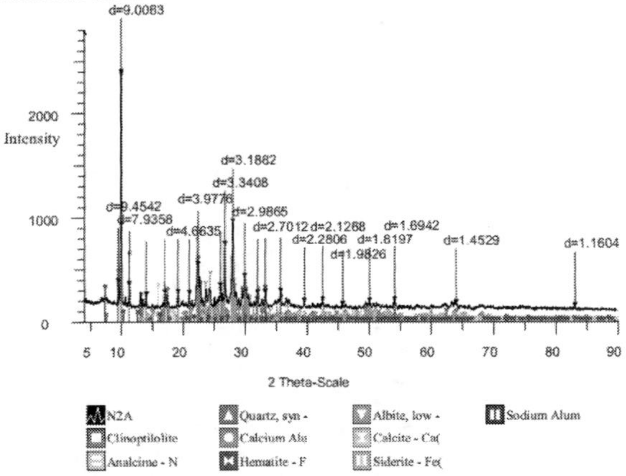

Fig. 1. X-ray of HKl-1 sample – obtained by treating natural zeolite of Shanhanay field by 1.75N hydrochloric acid.

As a result of acid activation there is an increase in the content of rock-forming minerals - clinoptilolite. Clinoptilolite phase in the sample HKl under the influence of 1.75N HCl compared with Kl source is increased by 20%.

Analysis of the IR spectra showed that natural Shankanay clinoptilolite sample is characterized by the following absorption bands: 465, 615, 780, 1060, 1635, 3460 cm^{-1} (Fig.2, spectrum 1). These characteristic absorption bands are present in the spectra of all the samples of zeolite (Fig.2, spectra 2 and 3). Here, spectrum bands 465, 615, 780 cm^{-1} result from internal and external vibrations alumo-silicate zeolite framework Si(Al). Most high intensity of the absorption band is observed in the frequency range 1060-1080 cm^{-1}, which correspond to the stretching vibrations of the Al-O-Si and Si-O-Si frame bonds.

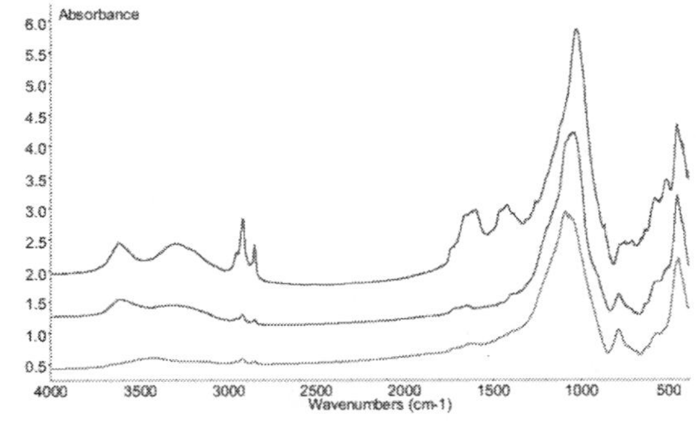

Fig. 2. IR spectra of the samples obtained by modifying the clinoptilolite 1 – Kl; 2 - HKl -1 ; 3 – HKl -1/Sal-1(1.75N HCl+10% HSal).

It is noticed that the spectra of the sample Kl treated by 1.75 N HCl, the relative intensities of the absorption bands are associated with the crystalline phases and within the tetrahedral vibrations do not undergo any significant changes. These facts suggest retaining the original crystal structure of the mineral. At the same time, it is observed the appearance of a new absorption band in the 796 cm^{-1}, also there is an increase in intensity are seen as a shoulder at 980 cm^{-1} band and an offset from 1095 to 1065 cm^{-1}.

The appearance of a new absorption band at 796 cm^{-1} corresponding to the stretching vibrations of Si-O-linkages indicates the formation of free silica and exit of aluminum to an exchange position. Absorption band at 796 cm^{-1} one can determine to the formation of new bonds Si-O-Si at break of Al-O-Si- bonds. Observed shifting of band from 1065 cm^{-1} to 1095 cm^{-1} due to the formation of new Si-O-Si-bonds due to release of Al from tetracoordinated position to an exchange position.

The above qualitative and quantitative X-ray and IR data indicate that when acidizing Kl undergoes deep dealumination without noticeable destruction of the structure of the mineral.

Thus, when used for treatment of relatively dilute acid clinoptilolite decationization occurs to form the hydrogen form, and simultaneously dealumination accompanied by partial amorphization of zeolite occurs.

Modification of the zeolite samples with an inorganic acid under optimal conditions enhance their catalytic properties (in the silicate modulus) sorption capacity and, but at the same time is not able to remove iron oxides from the zeolite.

Conclusin that in acid treatment clinoptilolit undergoes deep dealumination without appreciable destruction of the mineral structure is confirmed by X-ray obtained on the unit D8 Advance (Bruker), as well as IR data.

3.2 Modifying the natural zeolite with an organic acid

During activation of natural zeolite of Shankanay field by mineral acid there is easily removing cations of alkali and alkaline earth metals, and partly iron cations. However, the residual amount of iron in the composition of the zeolite, even under optimum conditions of acid activation, negatively effects on the catalyst (sorbent) performance, in particular on the hydrocarbons cracking process by catalyzing carbonization process from hydrocarbons.

To remove Fe^{3+} ions natural samples were treated by chelating agents: ethylenediaminetetraacetic acid (H4-EDTA, 10 and 25% solutions), sulfosalicylic acid (H-Sal, 10 and 25% solutions), citric acid (H-Lim, 10 and 25% solution), and disodium salt of H4-EDTA (Na2H2Y or Trilon B, 10% solution). All of them forms stable water-soluble complexes with iron ions, in particular the first choice for sulfosalicylic acid based on the ability of specific aromatic oxy acids, to form internal chelate compound.

Composition of samples obtained by direct modification of internal natural zeolite by some organic acids is shown in Table 2.

Table 2. The oxide composition of the samples modified with organic acids.

Oxide	Kl (natural)	Kl-H_4EDTA	Kl - HSal	Kl - HLim
Al_2O_3	12.1	12.8	9.6	14.1
SiO_2	68.0	74.0	74.0	71.0
Fe_2O_3	9.5	4.0	4.4	4.9
CaO	3.5	1.0	0.1	0.2
MgO	2.5	1.0	-	-
K_2O	2.3	1.5	-	-
Na_2O	2.1	1.0	2.8	2.0
SiO_2/Al_2O_3	9.6	9.9	13.1	8.6

When treating clinoptilolite with organic acids amount of alumina in its composition varies slightly, zeolite dealumination process is not compared with decationization of zeolite. This is evidenced by similar values of silicate source module (SiO_2/Al_2O_3=9.6) and modified (SiO_2/Al_2O_3= 9.9 - 13.1) zeolite samples. At the same time, the nature of the organic acid affects the silicate modulus of samples in different ways. Ethylenediaminetetraacetic acid regardless of the method does not change the value of the activation module silicate (SiO_2/Al_2O_3=9.6 and 9.9). Citric acid lowers the silica modulus clinoptilolite to 8.6, while sulfosalicylic acid increases it to 13.1.

With regard to the composition of oxides in the clinoptilolite, their relative contents in the dissolution under the action of organic acids are reduced. The amount of alkali metal oxides in the composition of the natural zeolite is reduced slightly at the same time oxides of alkaline-earth metals are removed almost completely. Content of iron oxide also decreases, and its residual amount is 4.0-5.0%.

It was revealed the effect of organic acids on the chemical structure and acidic properties of natural zeolite as compared with an inorganic acid is negligible. After organic acid modification natural zeolite by mainly decationization than dealumination processes occur. By modification of natural zeolite directly with organic acid the amount of undesired in catalytic processes iron oxide reduce only twice. Apparently, the organic acid do not reach iron ions, which are disposed within the crystal structure of clinoptilolite, but only interacts with the surface ions.

Among the organic acids used for decationization and dealumination as well as removal of iron ions from the clinoptilolite, using sulfosalicylic acid is more efficient.

3.3 Stepped modification of natural zeolite first with mineral and then by organic acids

Direct application of iron-chelate forming organic acids for modification of natural reduces the amount of iron oxide in the composition of the zeolite, but not changing the meaning of silica modulus of clinoptilolite.

In consequence, in order to obtain from Shankanay clinoptilolite catalysts with high acidity surface and low iron oxide content in the composition, natural zeolite is subjected to a stepwise modification of the mineral and the organic acids.

As the inorganic acid is taken 1.75N hydrochloric acid, as organic - sulfosalicylic acid (10 and 20% solutions) and ethylenediaminetetraacetic acid (10% solution).

Changes in the composition of the samples obtained by the two-stage modification of clinoptilolite are shown in Table 3.

Table 3. The oxide composition of the natural zeolite sample modified with two-stepped method.

Oxide	Kl (natural)	HKl-1/ Sal-1 (1.75N HCl; 10% HSal)	HKl-1/ HSal-2 (1.75N HCl; 20% HSal)	HKl-1/ H_4EDTA (1.75N HCl; 10% H_4EDTA)
Al_2O_3	12.1	7.2	7.8	10.7
SiO_2	68.0	82.0	77.0	78.3
Fe_2O_3	9.5	1.2	4.0	3.5
CaO	3.5	0.2	0.3	0.5
MgO	2.5	0.8	1.0	1.0
K_2O	2.3	1.5	2.0	2.0
Na_2O	2.1	-	1.5	1.0
SiO_2/Al_2O_3	9.6	19.4	16.8	12.6

When H_4EDTA is used in the second stage of acid activation, despite the significant decationization of zeolite, content of iron oxide is reduced to only 3.5%, while the value of the silica modulus of the sample remains low (12.6).

Application of another acid, 10% solution of sulfosalicylic acid, in the second step of modification dramatically reduces the iron oxide content of the zeolite in the sample to 1.2%. The extent of dealumination of the zeolite is also at a high level (SiO_2/Al_2O_3=19.4).

It should be noted that increasing the concentration of sulfosalicylic acid twice, from 10 to 20% in the second step did not improve the performance of the catalysts listed above.

In Figure 3 the effect of the nature of acids on change of the meaning of silica modulus is shown, also iron oxide content in the catalyst at the two-stage modification of natural zeolite is shown.

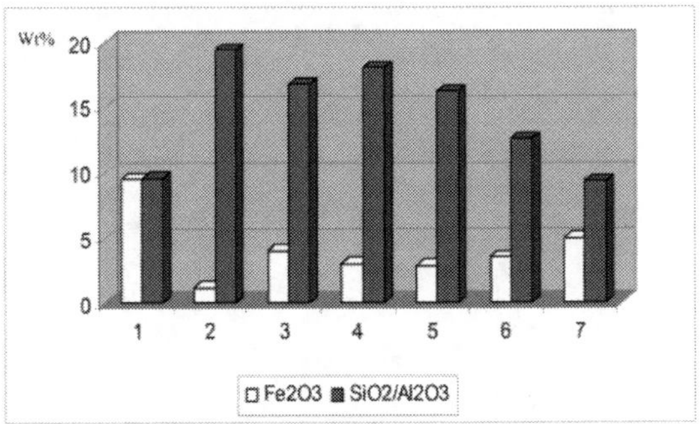

Fig. 3. The content of SiO_2/Al_2O_3 and Fe_2O_3 in the samples: 1 – Kl; 2 – HKl-1/HSal-1 (1.75N HCl-10% HSal); 3 – HKl-1/HSal-2 (1.75N HCl-20% HSal); 4 – HKl-1/HSal-2 (1.5N HCl-10% HSal), 98°C; 5- HKl-1/HSal-3 (1.0N HCl-10% HSal), 98°C 6 – HKl-1/H4EDTA (Soxhlet apparatus); 7 – HKl-1/Na_2H_2EDTA.

The crystallinity of the catalysts prepared by the two-stage process, first the inorganic and the organic acids were studied by X-ray and IR methods.

The IR spectra of the modified samples of natural zeolite first by inorganic HCl, then by organic acids HSal or H_4EDTA fully preserved absorption bands which are characteristic of the crystal structure of clinoptilolite. IR spectra of the sample of clinoptilolite for example, treated first solution 1.75N HCl, then a solution of 10% HSal characterized by absorption bands at 465, 581, 791, 1067-1085, 1636 cm^{-1} (see Figure 2, spectrum 3).

In X-ray diffraction patterns of zeolite - HKl-1/HSal-1, HKl-1/HSal-2 and HKl-1/H4EDTA observed a slight decrease in intensity of the reflection 8.92 Å and some decrease in sharpness of reflexes: 7.92; 6.65; 5.12; 3.98; 3.55; 2.97; 2.73 Å. Other responses, 7.92; 6.65; 5.12; 3.98; 3.55; 2.97; 2,73 Å, which are typical for clinoptilolite does not show noticeable changes when modification by two-step method.

Thus, for use as catalysts in petrochemical processes, by modifying first inorganic acid step, then the organic acid step decationized and dealuminated varying degrees Shankanay clinoptilolit samples created. It was found that samples prepared on the basis of zeolite previously decationized by 1.75 N hydrochloric acid with followed by treatment with 10% aqueous solution of sulfosalicylic acid (HSal) are characterized by the lowest content of iron oxide and the absence of destruction of the crystal structure of clinoptilolite.

3.4 Stepped modification of natural zeolite first with mineral and then with heteropolyacids

IR spectra (Fig.4) of the original and PW_{12}-HPA (or PMo_{12}-HPA) presented the following main absorption bands: 510, 594, 789, 867, 962, 1065, 1400, 1618, 2850, 2925, 3210, 3419 cm^{-1} which are characteristic for the connection of $H_3PW_{12}O_{40}{\bullet}6H_2O$ (or $H_3PMo_{12}O_{40}{\bullet}6H_2O$).

Comparative analysis of the IR spectroscopy the samples HKl-1, PW_{12}-HPA and PW_{12}-HPA/HKl-1 show that in these spectra characteristic absorption bands of clinoptilolite type zeolite are present: 465, 615, 780, 1060, 1635, 3460 cm^{-1}. The IR spectra of the acid activated Kl sample these absorption bands appear very clearly, i.e. when decationization and dealumination natural zeolite retains its crystalline structure.

As it is seen in Figure 4, modification with HPA acid activated Kl sample uniquely revealed in the IR spectra of the obtained zeolite samples. For example, when one apply PW12-HPA in the zeolite, on the IR spectra of the zeolite samples shift of some absorption bands characteristic of the zeolites and the appearance of new bands observed. Thus, a shift of the band at 1060 cm^{-1} to 1097 cm^{-1} and 1108 cm^{-1}. The absorption band of 780 cm^{-1} disappears, and at 800 cm^{-1} appears a new band, which is characteristic stretching vibration of Si-O-bond. Asymmetrical stretching vibrations at the end groups in the Si-O framework of clinoptilolite as indicated by the appearance and growth shoulder intensity at 1210 cm^{-1} in IR spectra. Given these changes can be concluded that in the zeolite structure Al-O-Si- linkages are broken and form Si-O-Si- linkages, aluminum enters the ion exchange position.

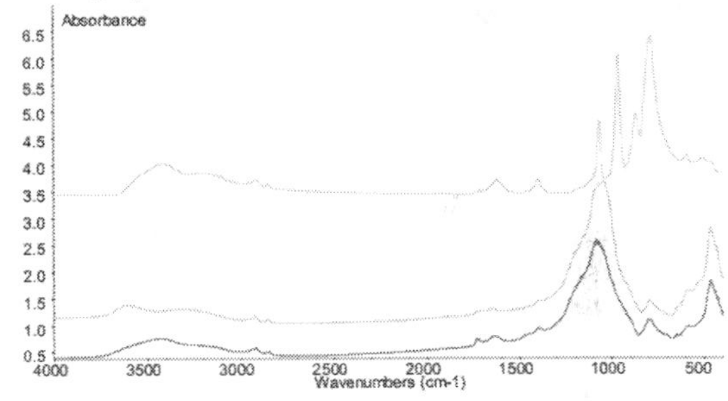

Fig. 4. IR spectra of samples 10% PW_{12}-HPA /HKl-1 and its components, where : a) IR spectra of the initial tungsten heteropolyacid- PW12-HPA ($H_3PW_{12}O_{40}{\bullet}6H_2O$), b) IR spectra of the HKl-1 sample, c) IR-spectra of sample 10% PW_{12}-HPA /HKl-1

X-ray of the samples also show a change in the structure of natural zeolite in the modification with HPA.

Thus, clearly revealed a decrease in the intensity of one of the main and the disappearance of some characteristic of the mineral clinoptilolite reflections in the diffraction patterns of samples 10% PMo_{12}-HPA/HKl-1 and 10% PW_{12}-HPA/HKl-1, which is apparently due to the partial amorphization of structures and changes in the composition of cationic surface of the zeolite.

These facts indicate the formation of a strong bond with the zeolite and HPA. It is believed that the substrate and the result achieved by calcination of catalyst PW_{12}-HPA translation in highly dispersed condition in which the particles appear catalyst PW_{12}-HPA specific adsorption and catalytic properties.

4 Conclusion

It can be concluded from the results that the modification of the sample of natural zeolite with various acids leads to changes in the composition and structure of clinoptilolite. Degrees of decationization and dealumination of clinoptilolite determined by the nature of the activating acid and bond strength of the cations in the mineral structure, which should be exchanged with hydrogen ions.

Indeed, all of the above factors have contributed to an increase in the catalytic activity of the cracking of paraffins [13,14]. In addition, an analysis of published data shows that ammonium salts of phosphorous-molibdenum heteropolyacid under certain conditions, can be used for the extraction of liquid radioactive waste of the lanthanides (Ce) and actinides (Am , Np) [16-19] . High selectivity of HPA salts with respect to radionuclides and high stability of natural ionizing radiation to the ion exchangers allowed using them effectively for the extraction of cesium from sea water, contaminated as a result of the accident at the Fukushima-I [16, 17].

The established relationship between the formations of catalytic systems based on the modification of natural zeolites and their catalytic activity offer great opportunities for targeted search for preparation of highly diverse hydrocarbon conversion catalysts.

References

1. Y.I. Isakov, H.M. Minachev Petrochemicals **3**, 291 (1990)
2. J.A.Gonzalez, N. Travieso, A. Rkoya, J.L. Seoane, Kinetics and Catalysis **32**(4), 907 (1991)
3. J. Rabo, *Zeolites Chemistry and Catalysis over zeolites* (Mir, Moscow, 1980)
4. Ch. Junwu, Acta petrol. sin. Petrol. Process. Sec. **20**(5), 1 (2004)
5. K.A. Mehdiyev, J. Chem. Problems. **2**, 324 (2007)
6. Z. Yuanzheng, Petrochem. Technol. **37**(8), 852 (2008)
7. I.V. Kozhevnikov, Phys. Chem. **56**(9), 1417 (1987)
8. T. Okuhara, Catalysts and Catalysis **45**, 32 (2003)
9. Y. Wang, J. Wang Petrochem. Technol. **32**, 453 (2003)
10. A.S. Kutergin, I.N. Kutergina Proc. of Ural State Technical University, **35**(5), 126 (2004)
11. A. Kilincarslan, S. J. Akyil, Radioanal Nucl. Chem. **264**(3), 541 (2005)
12. S. Akyil, M.J. Era,l Radioanal. Nucl. Chem., **266**(1), 89 (2005)
13. K.A. Kadirbekov Chem.journal of Kazakhstan **3**, 103 (2010)
14. S.R. Konuspayev, K.A. Kadirbekov, A.T Sarsekova, D.R. Nurbaeva, D.K. Zhambakin, Pet. Chem. **50**, 368 (2010)
15. K.A. Kadirbekov, Chem. J. of Kazakhstan **1**, 188 (2010)
16. A. Iwanade, N. Kasai, H. Hoshina, J. Radional. Nucl. Ch. **293**(2), 703 (2012)
17. Z. Huang, D. Zhang, Z. Tan, Adv. Mater. Res. **535**, 2191 (2012)
18. A.P. Gupta, P.K. Varshney, Syn. React. Inorg. Met. **28**(1), 105 (1998)
19. A.P. Gupta, S. Ikram, H. Agarwal, J. Indian. Chem. Soc. **80**, 57 (2003)

Preparation of Basalt Incorporated Polyethylene Composite with Enhanced Mechanical Properties for Various Applications

Pavel Bredikhin [1], *Yulia* Kadykova [2], *Igor* Burmistrov[1], *Tamara* Yudintseva[2,*], *Igor* Ilinykh[2], *Aleksandr* Kupava[2]

[1]Yuri Gagarin Saratov State Technical University, Saratov, Russia
[2]National University of Science and Technology MIS&S, Moscow, Russia

Abstract. The present article showed the possibility of increasing the complex of mechanical properties of polyolefins with dispersed mineral fillers obtained by fine grinding of basalt rocks via ball mill processing. The composites based on dispersed basalt, which were derived from Samara rock mass (Russia) with rare earth elements containing, were obtained by extrusion combining the binder and filler, followed by preparation injection-molded test samples. The study of mechanical properties of materials developed showed the possibility of a significant increase in strength characteristics of different types of polyethylene: the breaking stress at static bending for HDPE can be increasing more than 60% and the impact strength by more than 4 times. In addition the incorporation of the dispersed basalt also enhanced the thermal properties of the composites (the oxygen index of HDPE increases from 19 to 25%).

1 Introduction

During the processing of basalt rock, including the allocation of rare earths elements, may be formed the basalt powder waste. In this paper, one of the possible technologies of it secondary application was considered. Powdered and stone waste of ores processing containing rare earth can be used as additives in polymeric materials. Therefore in this article we have considered aspects of their influence on the mechanical properties of polymer matrix composites based on one of the most large-materials – polyethylene.

Polyolefins are generally the most commonly produced polymers, the amount of its production is more than tens of millions tons (35% of the world production of polymers) [1]. The most important representative of this class of polymers is polyethylene (PE).

PE is widely used due to its excellent properties such as: sufficiently high strength, resistance to corrosion, radiation, non-toxic and high dielectric properties [2]. However, the main disadvantage of PE is low modulus of elasticity and deformation under load, which limits its use. In addition, the incorporation of PE in various fillers significantly improves a complex of its physical and mechanical properties [2]. Moreover, the current economic climate requires not only the production of composites with a high characteristic complex, but also with a sufficiently low cost. In these reason various types of mineral fillers have become widespread: kaolin, talc, mica, chalk, silica, aerosil, titanium oxide, aluminum oxide, etc. [3-5]. Thus, most effective approach, to improve characteristics composite materials are incorporated in the use of inexpensive and efficient fillers, which include basalt and its derivatives thereof. Basalt is non-flammable, high temperatures resistant (900 °C), durable, resistant to mechanical stress, has high sound and heat insulating properties, biological stability, chemically non-reactive and non-radiation. The basalts are eco-friendly materials and non-toxic to humans and animals [6-7]. Due to these unique properties the basalt can be applied in various fields of engineering and technology. In the field of modification of polymer materials are used only basalt fiber and their derivatives. Moreover milled basalt as a dispersed filler has not been previously investigated. The aim of this work is to analyze the possibility of creating polymer composite materials based on polyethylene and dispersed basalt. In addition, study their physical-mechanical and chemical properties, due to the incorporation of polyethylene and dispersed basalt.

2 Materials and methods

As a binder low-density polyethylene (LDPE) and high-density polyethylene (HDPE) were used.

* Corresponding author: yudintsevatoma@gmail.com

Table 1. The chemical composition of the basalt.

Component	Content, wt%
Sodium oxide Na_2O	2.78
Potassium oxide K_2O	0.67
Calcium oxide CaO	9.35
Magnesium oxide MgO	15.86
Phosphorus oxide P_2O_5	0.5
Iron oxide Fe_2O_3	9.57
Silicon oxide SiO_2	49.88
Aluminum oxide Al_2O_3	10.09
Titanium dioxide TiO_2	1.08
Manganese oxide MnO	0.22
Loss on ignition	2.92

Derived from Samara rock mass basalt with rare earth elements containing (chemical composition thereof shown in Table 1 and Table 2) were used as a filter, in the present investigation.

Table 2. Content of rare earth and trace elements in the basalt.

Rare earth and trace elements	Content, g/tonne
Cr	780
Ni	630
Co	83
La	33.2
Ce	68.8
Pr	8.1
Nd	34.5
Sm	9.37
Eu	3.42
Gd	11.2
Tb	1.72
Dy	9.5
Ho	2.01
Er	5.5
Tm	0.77
Yb	4.22
Lu	0.64

Preparation of PE incorporated basalt composite: the basalt crushed in a ball mill 62ML-A (Mechanobr-Tekhnika, St. Petersbuig, Russia) for 15, 30 and 60 minutes at 50 rev/min. The ratio of the ground material weight to the grinding weight is 1:8. After grinding the unground particles product was sieved through 0.140 mm diameter. Shapes and sizes of the fillers, as well as the structure of the composites were analyzed using scanning electron microscope (SEM) Vega 3SB (Tescan, Czech Republic), at an accelerating voltage of 7 kV.

The composites were prepared by dispersing basalt in PE by using extruder granulator EPK 25x30 (Polyprom Kuznetsk, Russia). Test specimens were produced on an injection molding machine Haitian SA 900 (Haitian Plastics Machinery, China).

Melt flow index testing according to the ISO 1133 was made using Aflow Extrusion Plastometer 032295 (Zwick Roell, USA). Test conditions for HDPE: temperature – (190 ± 0,5) °C, load - 5 kg. Test conditions for LDPE: temperature – (190 ± 0.5) °C, load - 2.16 kg.

Breaking bending stress was determined according to ISO 178:2010; tensile stress at break and elongation at break were determined in accordance with ISO 527-1. Tests were carried out on a tensile testing machine INSTRON 5569 (Instron, USA). Brinell hardness was determined in accordance with ISO 2039/1-87 using HBE-3000A electronic Brinell Hardness Tester (Beijing United Test Co., Ltd., China). The impact strength was determined according to ISO 179-82 using a plastic tensile impact testing machine LCT - 50D (Beijing United Test Co., Ltd., China). The details of calculating of the test results are presented in section "Nomenclature list".

The composite materials were tested for flammability through oxygen index (EN ISO 4589-2).

2.1 Nomenclature list

Bending stress measurements:

$$\sigma_{sb} = \frac{3}{2}\frac{P*d}{b*h^2}$$

σ_{sb} – breaking stress at static bending, Pa; P – breaking load in static bending, N (it was measured with a tensile testing machine INSTRON 5569 according to ISO 178:2010); d – span, m (set 64 mm, according to ISO 178:2010); b – width of the sample, m; h – thickness of the sample, m (it was set recommended type of sample according to ISO 178:2010 with b=10.0±0.2 mm, h=4.0±0.2 mm).

Impact strength measurements:

$$a_i = \frac{P_i}{b*h}$$

a_i – impact strength, J/m^2; P_i – impact energy, J (it was measured with a plastic tensile impact testing machine LCT - 50D according to ISO 179-82; sample size: l=80.0±0.2 mm, b=10.0±0.2 mm, h=4.0±0.2 mm).

Tensile stress measurements:

$$\sigma_b = \frac{F}{b*h};$$

$$\varepsilon_b = \frac{F}{b*h}*100\%$$

σ_b – tensile stress at break, Pa; ε_b – elongation at break, %; F – tensile breaking load, N (it was measured with a tensile testing machine INSTRON 5569 according to ISO 527-1.)

Brinnel hardness measurements:

$$H_b = \frac{2*1839}{\pi*D*(D-\sqrt{D^2-d^2})}$$

H_b – Brinnel hardness, MPa; D – indenter diameter, m (it was measured with a electronic Brinell Hardness Tester HBE-3000A according to ISO 2039/1-87).

3 Results and discussion

The microscopic image observed through SEM of dispersed basalt is shown in Figure 1. The result

DOI: 10.1051/matecconf/20179600003

reveals that the particles are in irregular shape and size according to the milling time.

When the grinding is performed for 15 minutes the composition of the dispersed basalt predominantly presents the particles larger than 10 microns, with the number of smaller particles (Fig. 1). When grinding is carried for 30 minutes the formation of smaller sized particles around 3-6 microns were obtained. However, longer grinding for 60 minutes represents the formation of a large number of particles with sizes less than 1 micron, but the obtained powder included the particles with size about 5 microns and the aggregates of larger than 10 microns. Hereafter the obtained powders of the dispersed basalt are designated as follows: basalt after grinding for 15 minutes - DB15; basalt after grinding for 30 minutes - DB30; basalt after grinding for 60 minutes - DB60 (Fig. 1).

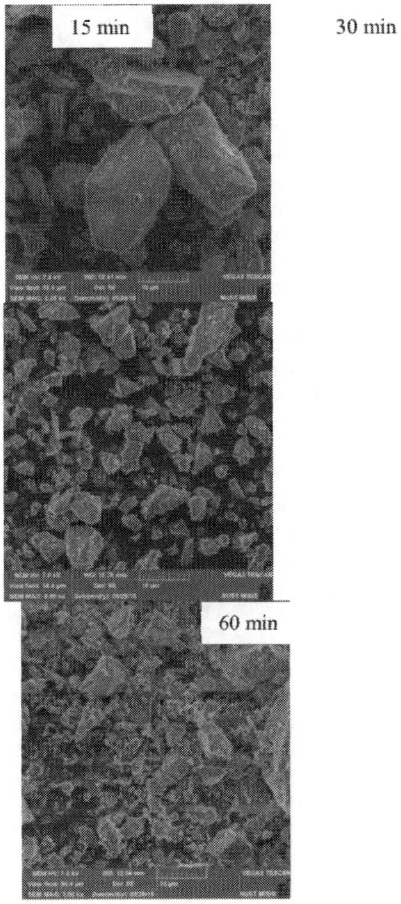

Fig. 1. Dependence particle size and shape of the dispersed basalt on grinding time with a planetary mill.

Evaluation of the rheological properties of the filled composites (Table 3) showed that with the increase in content of the basalt, the fluidity of the compositions decreased gradually when compared to unfilled systems. However, the obtained compositions have the melt flow rate big enough (more than 5 g / 10 min) for processing by the injection molding.

Table 3. Changing the melt flow index of the composition.

The composition, wt% per 100 wt% PE	Melt flow index, g/10 min
HDPE	23
HDPE+30 DB30	18
HDPE+40 DB30	16
HDPE+50 DB30	15
LDPE	5.5
LDPE+30 DB30	5.2
LDPE+40 DB30	5.1
LDPE+50 DB30	5.0

Study of physical and mechanical properties of developed polymer composite materials (PCM) showed that with the addition of the dispersed basalt there is an increase in resistance to static bending and impact loads (Figures 2-5). As demonstrated in figures, the best results achieved with the addition of 40 wt% of dispersed basalt. The extreme behaviour is typical for the strength properties dependency of the dispersed filler content in the polymer composite. This concentration dependence keeps for all particle sizes: DB15, DB30 and DB60.

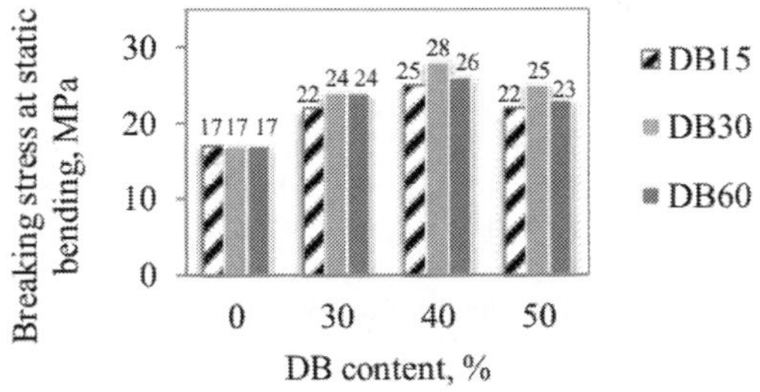

Fig. 2. Effect of content and particle size of the dispersed basalt on breaking stress at static bending HDPE.

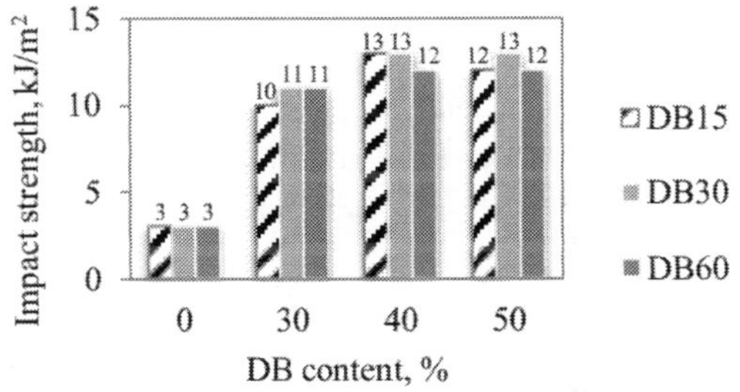

Fig. 3. Effect of content and particle size of the dispersed basalt on breaking stress at impact strength HDPE.

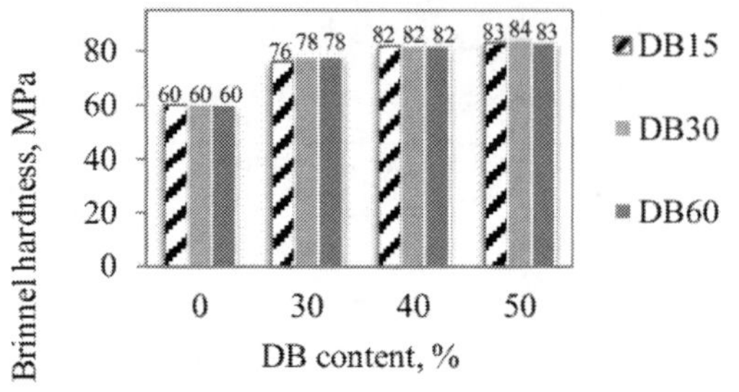

Fig. 4. Effect of content and particle size of the dispersed basalt on breaking stress at Brinnel hardness HDPE.

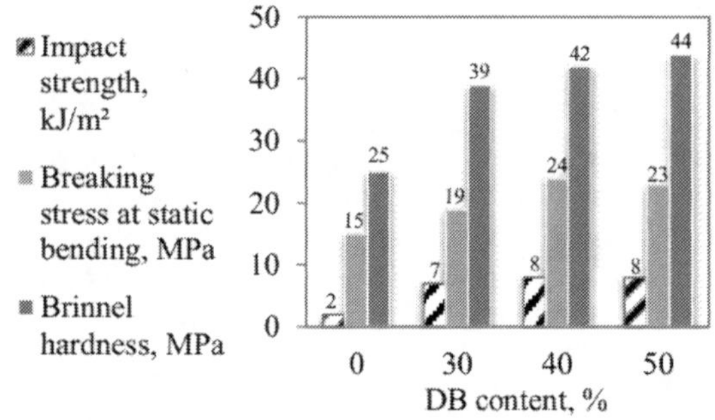

Fig. 5. Effect of dispersed basalt DB30 on the mechanical properties of LDPE m^2

By compare the reinforcing effect of various particle sizes of dispersed basalt the most effective method was for DB30. This might be due to the finer particles in comparison with DB15, higher homogeneous particle size distribution and the absence of aggregates in comparison to DB60.

It was found that the breaking stress at static bending for HDPE can be increasing more that 60% with incorporation 40 wt% of DB30 (Fig. 2). The impact strength for HDPE in this case can be increasing by more than 4 times (Fig. 3). This may be associated with significant energy losses in the dissemination of cracks in the highly filled polymer matrix.

It should be noted that the Brinell hardness insignificantly increasing (less 40%) when dispersed basalt incorporated into HDPE (Fig. 4).

With the introduction DB30 into LDPE matrix a significant increase of Brinell hardness (over 76%) was observed (Fig. 5). Other mechanical properties increase similarly to the HDPE (Fig. 5).

The addition of 40 wt% DB30 to the HDPE and LDPE the oxygen index increases from 19 to 25 and 24 vol % respectively. Basalt content double reduces the self-combustion compared to the unfilled PE (Fig. 6), and also reduced weight loss when ignited in air. All flammability indexes additively vary with basalt content, which is non-combustible material.

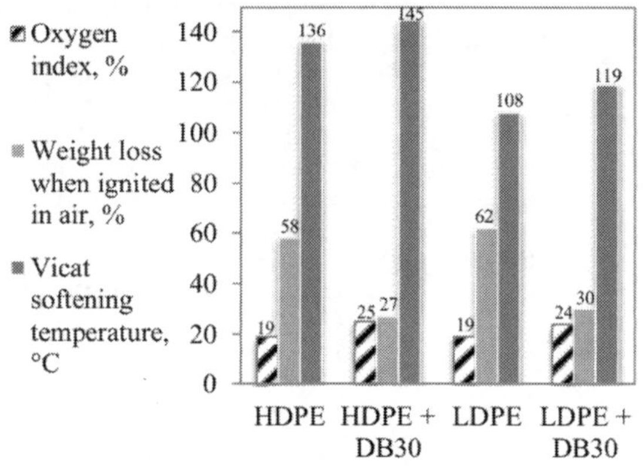

Fig. 6. Effect of DB30 on thermal resistance and combustibility of the filled polyethylene.

4 Conclusion

In this paper is shown the possibility of recycling of earth produce waste by introducing them into the polymer matrix composites. An important result shown in this study is the establishment of the possibility of increasing the strength characteristics of materials based on polyethylene with the addition of fine waste basalt products.

The efficiency of the use of dispersed basalt filling polyethylene was shown. The effect of dispersed basalt particle size on the indexes of physical and mechanical properties of HDPE and LDPE. Chosen optimal time of basalt grinding in a ball mill is analyzed to understand degree of fineness.

Enhanced physico-chemical and mechanical properties of the compositions filled with basalt, thus expanding the scope of basalt usage to create polymer composite materials for wide range of applications. In addition, the optimal content of dispersed basalt was found to be 40 wt%. In these case the breaking stress at static bending for HDPE has been increasing more than 60% and the impact strength by more than 4 times.

The influence of the basalt additives on indexes of polyethylene polymer composite materials flammability: oxygen index was analysed (it was increases from 19 to 25 %) based on self-combustion time, weight loss when ignited.

Dispersed basalt is extremely cheap material (in particular, it may be used waste of stone wool or fibers production), thus it can provide a significant increase of the strength of HDPE or LDPE, which is of great technical and economic importance.

This research was supported by Russian Ministry of Education and Science in frames of grants ID RFMEFI57814X0084.

References

1. M. Rahman, C. S. Brazel, Prog. Polym. Sci. **29**, 1233 (2004)

2. S. Ronca, *Chapter 10 – Polyethylene in Brydson's Plastics Materials (Eighth Edition)* (Elsevier Inc., 2017)), available at http://dx.doi.org/10.1016/B978-0-323-35824-8.00010-4
3. S. E. Artemenko, Yu. A. Kadykova, Fibre Chemistry **40**, 37 (2008)
4. I. N. Burmistrov, N. V. Shatrova, A. S. Mostovoy, I. N. Mazov, D. V. Kuznetsov, L. G. Panova, A. V. Gorokhovsky, A. G. Yudin, Polym. Eng. Sci. **54**, 2866 (2014)
5. D. S. Muratov, D. V. Kuznetsov, I. A. Ilinykh, I. N. Burmistrov, I. N. Mazov, Compos. Sci. Technol. **111**, 40 (2015)
6. S. E. Artemenko, Yu. A. Kadykova, Fibre Chemistry **40**, 490 (2008)
7. S. E. Artemenko, S. V. Arzamastsev, D. A. Shatunov, A. A. Vyazenkov, Fibre Chemistry **40**, 499 (2008)

Synthesis of Lithium Peroxide from Hydrogen Peroxide and Lithium Hydroxide in Aqueous-Organic Medium: Wasteless Technology

Roman Nefedov* and *Yury* Ferapontov

Tambov State Technical University, Tambov, Russia

Abstract. Lithium is a rare element, and widely used in manufacturing, electronics, medicine, and etc. One of the important lithium compounds is lithium peroxide. It is a component of regenerating products to protect the human respiratory system from damaging factors of chemical and biological nature. This paper describes the methods of obtaining the lithium peroxide. All industrial techniques for the synthesis of lithium peroxide are presented. A critical assessment of these techniques is given. A new wasteless synthesis technology of lithium peroxide from Hydrogen Peroxide and Lithium Hydroxide in Aqueous-Organic Medium is presented. This technology allows obtaining the pure product containing Li_2O_2 up to 94 %. The possibility of the solvent regeneration by the anhydrous lithium hydroxide has been shown. The yield of the lithium – 98 % and a significant reduction in solvent consumption per unit of finished product has been obtained.

1 Introduction

The inorganic peroxide compounds based on rare, radioactive, and especially alkali metals have a wide range of industrial applications (e.g., energy storage and oxygen source, etc.) [1, 2].

An identifying characteristic of inorganic peroxide compounds is the ability to produce oxygen in the active form by reaction with carbon dioxide and noxious impurities (human body waste products). That is why they are widely used in life-support systems (LSS), i.e. where a person is forced to stay and work in submarines, shelters, spaceships cabin that are isolated from the environment [4-6]. As a result, the development of new environmentally friendly technologies to obtain peroxide compounds of high purity on an industrial scale is a challenge by the present day. One of these compounds is lithium peroxide, which wasteless technology is described in this work.

There are a number of methods for producing lithium peroxide, each of which has its advantages and disadvantages. For example, lithium peroxide (Li_2O_2) was obtained in the early 20th century by Forkran [7] by the following method: aqueous-alcoholic solution containing 300 mL of 12% H_2O_2 and 450 mL of ethanol was added to 100 mL solution of lithium hydroxide containing 6.31 g LiOH. The resulting precipitate was dried for a long time over phosphorus pentoxide. The yield of a peroxide compound was 9 g or 72 % based on Li_2O_2. Under laboratory conditions a product containing 99.6 % Li_2O_2 [8] is obtained. The synthesis method Li_2O_2 proposed by Forkran has not undergone significant change by present day.

The use of ethanol is not a prerequisite for obtaining lithium peroxide, although it is beneficial in terms of that lithium peroxide contained in it unlike other organic substances practically insoluble (dissolved in methanol about 3.5 g/L Li_2O_2. Ethanol can be replaced by other organic solvents (methanol, pyridine, propanol, kenotic and others [12, 14]). One suggests considering the basic methods for the synthesis of hydrogen peroxide, lithium peroxide and lithium hydroxide with the use of organic solvents.

T.A. Dobrynina and S.Z. Makarov using the data on the solubility in the ternary system LiOH -H_2O_2 - H_2O and thermal stability of peroxohydrates and lithium peroxide hydrate [9, 10] developed two methods of synthesis of lithium peroxide, which were then implemented on the industrial scale [11, 12]. The study results of the thermal stability $Li_2O_2 \cdot 2H_2O \cdot H_2O_2$ and $Li_2O_2 \cdot H_2O$ demonstrated that these compounds in 85 – 100 ^{0}C can be easily decomposed to lithium peroxide. According to this, depending on the synthesis conditions, the process may run as the following equations of reactions:

$$2LiOH + 2H_2O_2 \rightarrow Li_2O_2 \cdot H_2O_2 \cdot 2H_2O \quad (1)$$

$$Li_2O_2 \cdot H_2O_2 \cdot 2H_2O \rightarrow Li_2O_2 + 3H_2O + 0.5O_2 \quad (2)$$

$$2LiOH + H_2O_2 \rightarrow Li_2O_2 \cdot H_2O + H_2O \quad (3)$$

$$Li_2O_2 \cdot H_2O \rightarrow Li_2O_2 + H_2O \quad (4)$$

Since the solubility of dehydrate peroxohydrate of lithium peroxide ($Li_2O_2{\cdot}H_2O_2{\cdot}2H_2O$) in aqueous solutions of hydrogen peroxide is 5-7 % based on the Li_2O as salting out agent in both cases, ethanol was used.

Synthesis of $Li_2O_2{\cdot}H_2O_2{\cdot}2H_2O$ was conducted by dissolving lithium hydroxide monohydrate under ice-cooling in 65-67% hydrogen peroxide. The resulting

* Corresponding author: kliorik@mail.ru

mixture was poured under continuous agitation into the ethanol, cooled to 0^0C. At the same time the formation of a white crystalline precipitate $Li_2O_2 \cdot H_2O_2 \cdot 2H_2O$ was observed. The starting materials were taken in the molar relation of $H_2O_2/LiOH = 3.9 / 1$. The resulting precipitate was separated from the mother solution by filtration, washed with absolute ethanol, dried in vacuum at 100 Pa. The yield based on lithium was 80-85 %. The final product contained 92-94 % of lithium hydroxide.

Preparation of the second lithium peroxide method carried out as follows. A suspension of lithium hydroxide in ethanol with vigorous stirring was brought to boil with 28 % hydrogen peroxide solution added after. The molar ratio of the components in the reaction mixture was $H_2O_2/LiOH = (0.75\text{-}1)/1$. The mixture was stirred under the boil for 1.2 hours. The resulting precipitate monohydrate lithium peroxide of a pale cream colour was separated from the mother liquor by filtration, washed with absolute ethanol and dried under vacuum at 80-120 ^{0}C. The yield on the lithium was 80-85 %. The final product contained 90 % of lithium peroxide.

In Li_2O_2 synthesis methods described above, there are two significant disadvantages – low yield for lithium and a high solvent consumption per unit of finished product. The reason for it is that the filtrate and the washing solution together are hard-separable mixture of water, solvent and lithium compound. That is why this mixture is removed from the process as waste. Thus, in the methods [12-14] about 15-20 % of the initial lithium compounds and all solvent are thrown out into the environment. It is not only environmentally harmful, but also unprofitable. Reuse of the solvent in the process can be held only by separating it from the water.

In patent [15] a method of lithium peroxide production with the help of n-propyl or isopropyl alcohol is described. A saturated aqueous solution of lithium hydroxide is prepared and then the required amount of 30 % hydrogen peroxide is added thereto. Obtained solution is cooled to 20 ^{0}C and n-propyl spirit is added with stirring. Thus, a compound of $Li_2O_2 \cdot H_2O_2 \cdot 2H_2O$ is obtained which is separated from the mother solution by filtration and dried in vacuum. For reuse (regeneration) of the solvent precipitation from filtrate of lithium as carbonate is carried out by the injection in filtrate anhydrous potassium carbonate. Precipitated lithium carbonate is separated from the filtrate and the solvent is returned to the stage of the reaction of lithium hydroxide and hydrogen peroxide.

However, the injection of potassium carbonate in the filtrate leads to accumulation therein of ions of potassium and carbonate – anions, which leads to contamination of the final product - lithium peroxide by these ions and hence reduces the mass fraction of a base substance. In addition, since the filtrate contains excess water that cannot be completely removed by the injection of potassium carbonate, returned to a cycle solvent also contains water which leads firstly to forming of a precipitate of peroxide lithium compound containing a large amount of water and secondly adding of extra quantity of anhydrous solvent is required in order to intensify the withdrawal of water from the reaction zone. In addition, the filtrate contains water excess, which cannot be completely removed by the introduction of potassium carbonate. The recycled solvent also contains the water. In the result, a precipitate of lithium peroxide containing a lot of water. For the intensifying the withdrawal of water from the reaction zone, an addition of further anhydrous solvent is needed. The lithium carbonate, formed from the filtrate processing, cannot be returned to the production cycle without further processing into the lithium hydroxide, and it is removed as a waste. As a result, from cycle to cycle, a most expensive component (the lithium) is lost during synthesis.

Hence, though this method reduces the consumption of dissolvent by bringing it back to the cycle, it does not provide a sufficiently high yield of lithium and thus it is inefficient. Industrial processes for obtaining lithium peroxide are known where the organic solvents are not used completely or only in the final stages of chemical-technological process. The "Lithium Corp. of America" company developed several options for the lithium peroxide of industrial production [16, 17].

Lithium peroxide can be received in ample quantity through lithium alkoxides as well, because the solubility of lithium methoxide in methanol is 90 g / L [12].

These methods along with ones mentioned above have a significant disadvantage-low yield of lithium (under 85%).

Other methods of preparing lithium peroxide - the oxidation of metallic lithium by oxygen, hydroxide, oxide and lithium amalgam [12, 17], the reaction of lithium nitrate with sodium superoxide in liquid ammonia [17] are not of interest for the development of new industrial technologies, leading to the formation of products containing 30-75% Li_2O_2 or they remain laboratory methods [18].

Thus, the development of a new obtaining industrial technologies peroxide compounds of high purity is a challenge. Current study develops a cost- and resource-effective, wasteless synthesis technology of Lithium Peroxide.

2 Methods and Materials

Developed environmentally friendly technology for producing lithium peroxide of lithium hydroxide and hydrogen peroxide in-ethanol in the environment is implemented in process flow diagram (Fig. 1) and includes the following technological stages and operations.

The reactor 5 from the measuring device 1 received a certain amount of ethanol and lithium hydroxide monohydrate from the feeder 3, where they were mixed by a mechanical stirrer. The obtained reaction mixture was heated to boiling direct steam entering the reactor casing 1. After that the temperature in the reaction zone was reduced to 75 ^{0}C

This method by the increase of the center of crystallization allows receiving a finer and consequently more reactive product, which is essential for its subsequent use in LSS systems. Ethanol vapors were entrapped in the reflux condenser 4 and returned back to the reactor 1. Then, 30 % hydrogen peroxide was added from the measuring device 2. The temperature in the reaction zone was monitored by a contact thermocouple and maintained at a constant level..

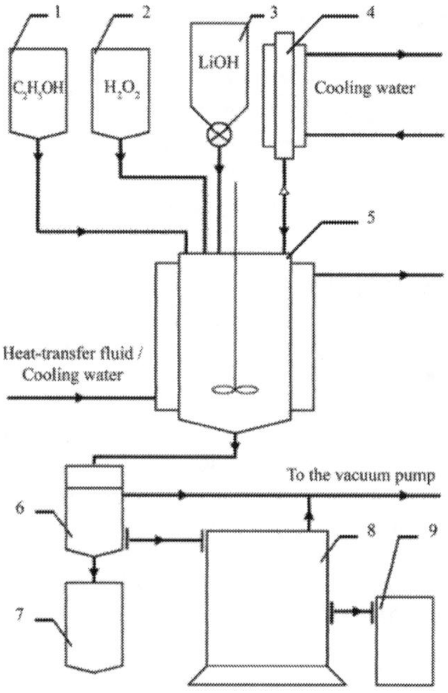

Fig. 1. Technological scheme of receiving peroxide of lithium: 1,2 – measuring tanks, 3 – star feeder, 4 – reflux condenser, 5 – reactor, 6 – suction filter, 7 – capacity, 8 – vacuum drying cabinet, 9 – container.

The precipitate formed $Li_2O_2{\cdot}H_2O_2{\cdot}2H_2O$ was filtered at suction-filter 6, washed with two portions of ethanol. After that it was dried in a vacuum oven 8 for 30-40 minutes at a temperature of 35 to 40 ^{0}C on griddles, made of stainless steel. Then, the temperature was increased to 125-135 ^{0}C and the product was dried for 8–9 h. The residual pressure in the vacuum-oven was 4.5 kPa. The finished product was collected in a container 9.

3 Results and discussion

X-ray phase analysis (XRD) of the lithium peroxide samples has been carried out. Radiographs of the samples were recorded on a DRON-6 with the filtered Cu Ka- radiation (l = 0.154051 nm). Data with XRD results and calculated based on diffraction patterns values of interplanar distance (A^o) are shown in Fig. 2. XRD data, as well as the results of chemical analysis of these samples showed no water in their composition. So, dehydration process runs fully under conditions of experiments.

The composition of the resulting precipitate ($Li_2O_2{\cdot}H_2O_2{\cdot}2H_2O$) was determined by chemical analysis and DTA method. The obtained data indicates that in the described conditions the reaction route between the hydrogen peroxide and lithium hydroxide-guide is expressed by equation 1. Lithium yield was 91 %, the final product contained 93.4 % of lithium peroxide.

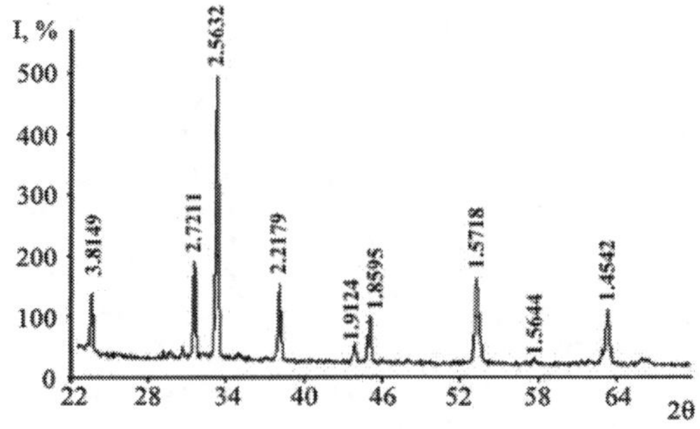

Fig. 2. The diffraction pattern of the sample obtained peroxide lithium.

In order to use ethanol in the next cycle, in the mother solution collected in container 7, the amount of anhydrous lithium hydroxide calculated according to the equation of reaction (5) was added, which reacts with water to form lithium hydroxide monohydrate.

$$LiOH + H_2O \rightarrow LiOH{\cdot}H_2O \downarrow \quad (5)$$

The precipitate $LiOH{\cdot}H_2O$ was filtered on a suction-filter 6 and processed its dehydration in a vacuum oven at a temperature of 150-170 ^{0}C. After dehydration anhydrous lithium hydroxide weight (taking into account technological and logical losses) was not changed. The missing amount of ethanol was added to solvent remaining after regeneration, and returned to the first stage of the process in the reactor 1, so the regenerated solvent was mixed with a new portion of the mono-hydrate of lithium hydroxide and subsequent operations were carried out to obtain lithium peroxide. The yield on the lithium increased to 98%. The final product contained 93.4% of lithium peroxide.

There is main advantage using of the mother solution of anhydrous lithium hydroxide for regeneration: secondary ions are not entered into mother solution. It means the absence of contamination of the desired product. To identify the optimal conditions of this process, a series of experiments with different ratios of the reactants and temperature were carried out. All data are presented in Table 1.

The data in Table 1 shows that the process is best performed at a molar ratio of the starting components $LiOH/H_2O_2$ = 1/(1.10-1.15). In this case, a final product with a maximum content of basic substance is received. Further increase (as well as decrease) the amount of added hydrogen peroxide does not increase the content of the basic substance in the product synthesis. The preferred temperature of the process flow is 75 ^{0}C. Reducing the temperature in the reaction zone also reduces the basic substance in the product synthesis. Using lithium hydroxide for regeneration of solvent allows reducing consumption of ethanol for 5 cycles 67.5 liters to 2.5 liters in the preparation of 3.3 kg of lithium peroxide and thus increasing the yield of lithium from 91.5% to 96.7% without deteriorating the quality of the product.

Table 1. The data of experiments on identification of optimum conditions. T, ^{0}C – temperature in a reaction zone; H_2O_2 (g) / LiOH (g) / EtOH (L) – the cconsumption of reagents for 5 cycles; m, g – the amount of peroxide of lithium received for 5 cycles; α, % –the yield on lithium for 5 cycles; ω, % – mass fraction of the main substance.

No	Molar ratio $LiOH/H_2O_2$	T,^{0}C	H_2O_2(g) LiOH(g) EtOH (L)	m, g	α, %	ω, %
1	1/1.10	75	5565.1 3571.5 2.5	3306.1	96.6	93.4
2	1/1.15	75	5818.1 3571.5 2.5	3309.5	96.7	94,0
3	1/1.20	75	6071 3571.5 2.5	3306.1	96.6	93.4
4	1	75	5059.2 3571.5 2.5	3280.8	95.8	86.0
5	1/1.10	75	5565.1 3571.5 2.5	3312.9	96.8	94.0
6	1/1.15	70	5818.1 3571.5 2.5	3281.4	95.9	88.2
7	1/1.15	75	5818.1 3571.5 67.5	3131.5	91.5	94.0

NOTE. The amount of consumable hydrogen peroxide (H_2O) is given in terms of 100% substance. The amount of consumable lithium hydroxide monohydrate ($LiOH \cdot H_2O$) is given in terms of anhydrous lithium hydroxide (LiOH). The following table shows the results of obtaining lithium peroxide for 5 cycles. In this case, one cycle is 0.5 liters of solvent consumption and for five cycles - 2.5 liters, while the solvent-free regeneration (Example 7) solvent consumption for the same five cycles was 67.5 L.

4 Conclusion

A new environmentally friendly synthesis technology of lithium peroxide is presented. The technology is based on interaction of hydrogen peroxide solution with a hydroxide of lithium in ethanol followed by vacuum - thermal dehydration of the intermediate adduct dehydrate peroxohydrate peroxide lithium ($Li_2O_2{\cdot}H_2O_2{\cdot}2H_2O$), allowing to obtain the final product containing 94% of the main substance.

To regenerate the solvent anhydrous lithium hydroxide was used which made possible to increase the yield of lithium from 91 to 98% and to reduce solvent consumption significantly per unit of final product without introducing extraneous ions into the final product.

It has been established experimentally that the optimum process conditions for interaction of the hydrogen peroxide solution with lithium hydroxide in ethanol T = 75^{0}C, and the molar ratio of the starting reactants $LiOH/H_2O_2$ = 1/(1.10-1.15).

Thus, the paper demonstrates, that the technology is a cost- and resource-effective. The main advantage of this technology - no waste, and it can be used on an industrial scale.

The study was performed on the expenses of a grant from the Russian Science Foundation (project #15-19-10028).

References

1. M.D. Radin, J.F. Rodriguez, F. Tian, D.J. Siegel, J. Amer. Chem. Society **134**, 2 (2012)
2. N. Deng, W. Wang, G. Yang, Y. Qiu, RSC Advances **126**, 5 (2015)
3. T. Gladysheva, N.F. Gladyshev, S. I. Dvoretsky *Nanocrystalline Regenerative Products. Synthesis, Properties, Application* (Spectr, Moscow, 2014, in Russian)
4. J. Presti, H. Wallman, A. Petroctlli, Undersea Technol. **8**, 20 (1967)
5. G.I. Voronin A.I. Polivoda, *Sustenance Space Crews* (Mashinostroenie, Moscow, 1967, in Russian)
6. A. Petroctlli, A. Capolesto, Aerospase Med. **31**, 440 (1964)
7. Forcrand R. De. - C.r. **130**, 1465 (1900)
8. G. Brauer, *Guide to Inorganic Synthesis* (Mir, Moscow, 1985, in Russian)
9. S.Z. Makarov, T.A. Dobrynina, News of the Academy of Sci.: Chem. Sci. **3**, 411 (1955) (in Russian - Izvestya Academii Nauk SSSR OKhN).
10. S.Z. Makarov, T.A. Dobrynina, News of the Academy of Sci.: Chem. Sci. **3**, 294 (1956) (in Russian - Izvestya Academii Nauk SSSR OKhN)
11. R.O. Bach, W.W. Boardman, U.S. Patent No. 3134646, (26 May 1964)
12. T.A. Dobrynina, *Lithium Peroxide* (Nauka, Moscow, 1964, in Russian)
13. H.H. Strater, M. Heights, U.S. Patent No. 2,962,358 (29 November 1964)
14. R.O. Bach, W.W Boardman, U.S. Patent No. 3,147,075 (1 September 1964)
15. E.L. Klebba, U.S. Patent No. 3,212,850 (19 October 1965)
16. I.I. Vol'nov, *Peroxide compounds of Alkali Metals* (Nauka, Moscow, 1980, in Russian)
17. P. Winternitz, U.S. Patent No. 2,488,485 (15 November 1949)
18. Yu.A. Ferapontov, S.I. Simanenkov, N.F. Gladyshev, B.V. Putin, RF Patent No. 2,193,522 (27 November 2002)

Refining of Rough Indium by Method of Reactionary Electrolysis

Bibisara Burkitbayeva, *Akmaral* Argimbayeva, *Gulmira* Rakhymbay*, *Khaisa* Avchukir, *Khaidar* Tassibekov, and *Mikhail* Nauryzbayev

Al-Farabi Kazakh National University, Center of Physico-Chemical Methods of Research and Analysis, Almaty, Kazakhstan

Abstract. Electro-refining of rough indium by method of reactionary electrolysis from the chloride solutions containing $InCl_3$ was carried out. The research of ionization of rough and modified indium anodes showed a decrease in an overvoltage of dissolution of indium when using the anodes containing bismuth. The values of the activation energy of the ionization stage of rough to 7.3 kJ / mole and the indium anodes modified by bismuth with 5- and 10-fold excess of depressor in relation to the total content of impurities, amounted to 5.6 and 3.7 kJ / mole, respectively. Results of the analysis of the refined indium by ICP-OES and ICP-MS methods showed that use of bismuth as a depressor additive allowed reducing impurity (Zn, Pb, Cd, Cu, Fe, Ni, As, Tl) level to 1.1 ppm.

1 Introduction

Indium is a rare and trace elements, with a number of unique properties. For use in a variety of industries: production of screens LCD monitors, TVs, and LED, fiber optics and electronics indium of high purity with a total impurity content of 10-0.01ppm is required [1].

In the works [2-4] devoted to refining of indium in various electrolytes on solid electrodes the multistage techniques including chemical refining, vacuum still and electro-refining are presented. When using of the combined methods from chloride and sulphate electrolytes optimum conditions of purification of indium were established and metal of brand 6N [5] is received.

The authors of papers [6, 7] developed a highly effective method of reactionary electrolysis with electrolyte pre-refining. This method is based on the principle of binding the impurities, contained in the purified metal in electrochemically inactive compounds by antimony additives.

In this paper, for purification of rough indium the method of reactionary electrolysis was used, with high productivity, the ability to process automation, ease of hardware design and the lack of multi-stage. This method allows reducing the total impurity content in the reduced indium.

Research objective is to receive the cathode precipitations of indium of high purity with minimum energy consumption, using the method of electrolysis reaction, in which bismuth is used as the depressor additive for the first time. In comparison with antimony, bismuth forms the stronger electrochemical inactive inter-metallides with the difficult deleted impurity metals.

2 Experimental part

Preparation of the modified anodes was carried out in the cylindrical heat-resistant reactor made of quartz and included two stages: preparation of a fusible alloy (ligature) of $In+In_2Bi$ and receiving anodes with various content of bismuth.

The ligature of a bismuth was prepared by an alloy of indium of the In-2 brand (78%) with a bismuth (purity of 99.999%, Sigma Aldrich) (22%) at 270^0C under a paraffin layer when hashing within one hour which was used further for manufacture of the modified anodes of indium.

Rough indium of the IN-2 brand contains 0.01 wt. % of the impurity metals. Two modified anodes were prepared: A2 – with 5-fold, A3 – with 10-fold excess of bismuth in relation to the total content of metals-impurity. Proceeding from the calculated necessary content of bismuth, the quantity of a ligature was defined and alloyed with rough indium at a temperature of 250^0C under a paraffin layer within one hour with continuously mixing a melt.

Electro-oxidation of indium is investigated by method of the linear voltammetry in a temperature-controlled three-electrode cell with a capacity of 50 ml, with use of a potentsiostat-galvanostat of Autolab PGSTAT 302N. The working electrodes were rough (A1) and the indium electrodes modified by bismuth (A2, A3) with an area of 0.196 cm^2. The auxiliary electrode was a platinum plate with an area of 12.96 cm^2, silver chloride electrode served as the reference electrode.

All experiments were made at a temperature of 25^0C, with an accuracy of ±0.1. Polarization resistance measurements were carried out at -10 mV to +10mV vs.

* Corresponding author: gulmirarahymbay@gmail.com

DOI: 10.1051/matecconf/20179600005

E_{OCP} at scan rate of 5 mV/s. Polarization curves recorded again with a scan rate of 5 mV/s in the range of – 300 mV to +300 mV vs. E_{OCP}.

Chloride solution containing 0.5 M $InCl_3$+2.0 M NaCl (pH = 1.5) was used as the electrolyte during the electrochemical study of oxidation of indium and electrolysis. Supporting electrolyte prepared from sodium chloride received by a triple recrystallization of chemically pure salt (99.9%). The pH 1.5 of solution was created by acidification with hydrochloric acid.

Use of chloride electrolytes with the content of indium salt of 0.5 mole/l allowed obtaining qualitative precipitates of indium with a large current output. The choice of 2.0 M solution of sodium chloride as a blank electrolyte is caused by its high conductivity and optimum viscosity. As it follows from work [4] the variation of concentration of Sodium chloride does not influence on the size of a current output, however an increase in its contents considerably reduces a voltage drop in an electrolysis bath.

Electrochemical cell for the electrolysis consisted of anode (A1, A2 or A3) and titanium cathode (BT1-0) with an area of 1 cm^2. Preparation of the anode and the cathode surfaces was carried out mechanically using the micron cloth and rinsing with bidistilled water before each measurement. To assess the reproducibility, each measurement was performed at least three times. Refining of indium using rough and modified anodes was performed in the galvanostatic mode at room temperature at a current density of 300 A / m^2. Effective inter-electrode distance was 2.0 cm. After electrolysis the cathode precipitate thoroughly was washed with redistilled water, and dried in a desiccator for a day.

Samples were prepared by dissolving 0.1 g of indium precipitated in 2.5 ml of nitric acid (OFS) and bringing to the mark with double-distilled water in a volumetric flask of 25 ml. The composition of the samples of refined indium was determined by optical emission spectrometry (ICP-OES) and mass spectrometry (ICP-MS) with inductively coupled plasma on Optima 8000 (Perkin Elmer Company) and ICP-MS Agilent-75a, respectively.

3 Results and discussion

3.1 Anodic dissolution of rough and modified indium anodes.

One of stages of electrochemical refining of metal is its oxidation. In [8, 9] oxidation of indium in sulphate and chloride electrolytes at various temperatures to determine the optimal conditions for the electrochemical refining of rough indium was investigated. The anodic dissolution of indium in chloride electrolytes proceeds with a lower overvoltage, than in the perchlorate and sulphate, that is explained by an activating effect of chloride ions and the higher stability of chloride complexes of indium.

In this work the anode behavior of rough and the modified indium anodes containing 5 and 10-fold excess of bismuth relative to the content of metal impurities is studied. Insufficient bismuth content in the anode may result in its preferential interaction with indium and incomplete binding of impurities, especially at their small contents. Excessive bismuth content can lead to contamination by bismuth of cathode precipitates. The use of bismuth is due to the formation of strong bonds Bi-Tl and Bi-Zn, which is evident from a comparison of the values of change of the Gibbs free energy of formation of bonds Bi-Tl, Bi-Zn and In-Bi.

Selection of a ligatures composition according to the phase diagram of the In-Bi system, is caused by its low melting point (72^{0}C) and high indium content that improves the distribution of the inter-metallide In_2Bi in the volume of melt [10].

Electrochemical disolution of indium anodes is investigated by the potentio-dynamic method which allowed to determine kinetic parameters of process (Fig. 1).

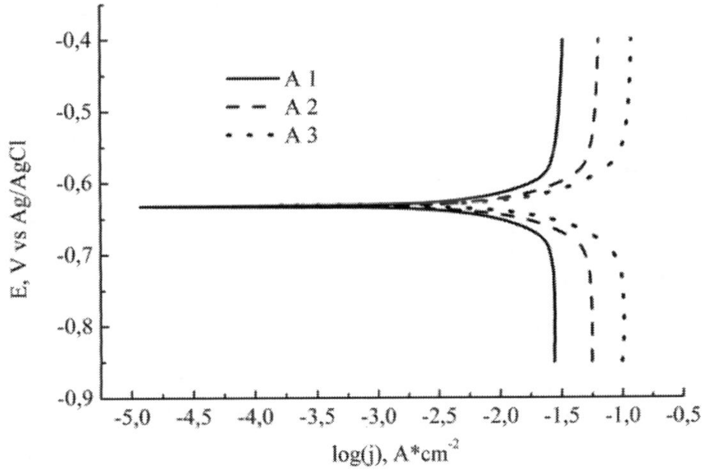

Fig. 1. Cathodic and anodic polarization curves for various anodes in the solution of 0.5 M $InCl_3$ + 2.0 M NaCl, υ = 5 mV/s, T = 25 °C: A1 - rough indium, A2 - modified indium with 5 fold, A3 – 10 fold excess of Bi relative to the total content of metal impurities.

As seen in Figure 1, an increase in the content of a depressant in the anode leads to the growth of indium ionization current without changing the equilibrium potential. From polarization curves the polarization resistance, exchange current and Tafel's constant b_a, which according to the Battler-Folmer equation is directly proportional to a charge transport coefficient (1-α) (Table 1), were calculated.

Table 1. onization parameters of rough and modified indium anodes.

Anode	j_o, mA/cm^2	b_a, mV/dec	R_p, Ω*cm^2
A1	19.98	189	1.84
A2	67.86	231	1.11
A3	103.24	293	0.72

Note: A1 - rough indium, A2 - modified indium with 5-fold, A3 – 10-fold excess of bismuth (relative to the total content of metal impurities).

With an increase in the content of bismuth in the anode a decrease in the polarization resistance occurs. This leads to an increase in the exchange current. This apparently is due to the activating effect of depressor additives that cause a decrease of the energy barrier of electron transfer, which is characterized by the coefficient of electron transfer proportional to b_a. The values of the activation energy of process of ionization of rough (A1) and modified (A2, A3) of indium anodes, received from the analysis of the dependence of values of the logarithm of their exchange currents on temperature (Fig 2.), were 7.3; 5.6 and 3.7 kJ /mole, respectively.

Values of activation energy testify to the diffusion nature of process of ionization. The obtained data indicate a decrease in an overvoltage (over-potential) during the oxidation of indium in the modified electrodes in the following row: $\eta_{(A1)} > \eta_{(A2)} > \eta_{(A3)}$.

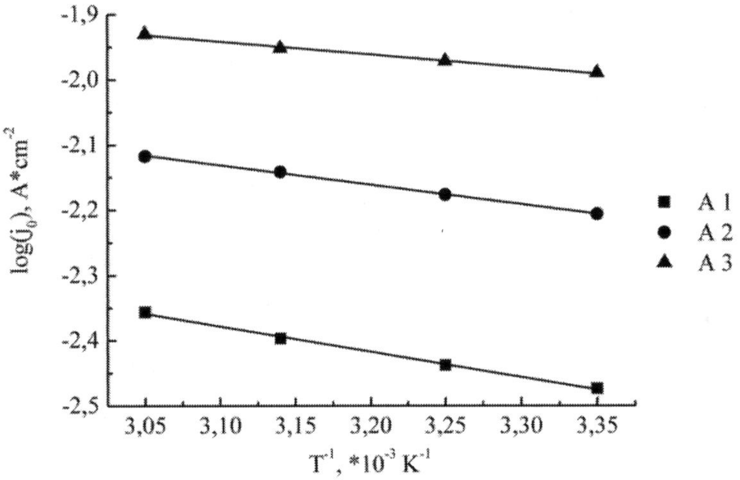

Fig. 2. Dependence of the exchange current density logarithm from inverse temperature values for different anodes in the solution of 0.5 M $InCl_3$ + 2.0 M NaCl, υ = 0.05 V/s: A1 - rough indium, A2 - modified indium with 5 fold, A3 – 10 fold excess of Bi relative to the total content of metal impurities.

3.2 Electro-refining of indium

The character of the electrode processes during the electrolysis depends on many factors: the electrode material, the composition of electrolyte, electrolysis mode (current density, temperature and inter-electrode distance). Electrolysis was carried out with rough and modified indium anodes with different content of bismuth.

From the measured values of voltage on the electrolytic cell, the specific energy consumption has been calculated for 1 t of cathode precipitates (Table 2).

The received results showed that modifying of the anode by bismuth at electrochemical refining of indium contributes to maintaining of high yield of indium on current, the voltage drop in the electrolytic cell and consequently a decrease in specific energy consumption.

Table 2. Energy characteristics of electrolytic refining of rough and modified indium anodes.

Anode	Voltage (U), V	Current efficiency, %	Specific energy consumption (SEC), kW·h·t^{-1}
A1	0.62	99.3	430
A2	0.52	98.3	371
A3	0.55	99.9	386

Note: A1 - rough indium, A2 - modified indium with 5-fold, A3 – 10-fold excess of bismuth (relative to the total content of metal impurities).

Table 3. Results of ICP-OES analysis of the composition of rough indium and cathode precipitates of indium.

Me	Composition of rough indium, wt.*10^{-3} %	Composition of cathode precipitates, wt.*10^{-5} %		
		A1	A2	A3
Cd	1.00	3.23	0.62	1.77
Cu	1.00	7.69	-	-
Fe	1.00	-	-	-
Ni	0.50	4.05	0.91	0.70
Pb	1.00	83.50	0.65	0.97
Sn	1.00	-	1.63	-
Tl	1.00	70.50	1.67	6.05
Zn	3.00	2.91	2.19	0.19
In	99.99050	99.99825	99.99989	99.99940

Note: A1 - rough indium, A2 - modified indium with 5 fold, A3 – 10 fold excess of Bi relative to the total content of metal impurities.

Table 4. Results of ICP-MS analysis of the composition of rough indium and indium cathode precipitates.

Me	Composition of rough indium, wt.*10^{-3} %	Composition of cathode precipitates, wt.*10^{-5} %		
		A1	A2	A3
Cd	1.00	1.7	1.4	1.77
Cu	1.00	5.13	3.72	6.05
Fe	1.00	-	-	-
Ni	0.50	3.95	-	2.52
Pb	1.00	87.5	-	32.8
As	1.00	-	1.63	-
Tl	1.00	69.5	1.67	6.05
Zn	3.00	-	0.93	-
In	99.99050	99.99832	99.99990	99.99951

Note: A1 - rough indium, A2 - modified indium with 5 fold, A3 – 10 fold excess of Bi relative to the total content of metal impurities.

Studying of anodic dissolution of indium anodes showed a decrease of values of activation energy and polarization resistance in case of use of depressor additive. This is a proof of decrease in energy costs at refining of the modified rough indium.

The results of ICP -OES and ICP- MS analysis of cathode precipitates of indium at the refining of rough and modified indium anodes are presented in Table 3 and Table 4.

As appears from the above-presented results (Table 3 and Table 4) the content of the main metal, after cleaning of the anode of rough indium, made 99.998% that means an increase in purity of indium. In case of refining of the anode modified with 5-fold excess of bismuth, purity of cathode precipitates considerably increases (99.99989%), that is the basis of possible use of bismuth as depressor additive.

4 Conclusion

The indium refinement by a reactionary electrolysis in the presence of a bismuth as the promoting additive which allowed reducing activation energy of a stage of ionization of indium from 7.3 kJ / mole for the anode from rough metal to 3.7 kJ / mole - for the modified electrode was carried out. Bismuth additive in anode material leads to a voltage drop in an electrolytic bath on 100 mV and to decrease of energy consumption.

Results of the analysis of the refined indium demonstrate increase in purity of metal by 90 times.

References

1. A.G. Morachevskii, Rus. Appl. Chem. **79**(3), 516 (2006)
2. L.A. Kazanbaev, P.A. Kozlov, V.L. Kubasov, *Indium Production technology* (Ore and Metals, Moscow, 2004)
3. M.S. Lee, K.Y. Sohn, J. Mater. Sci. **38**(24), 4843 (2003)
4. J. C. Park, Bull. Korean Chem. Soc. **32**(10) 3796 (2011)
5. Z.-H. Zhou, J.-M. Ruan, H.-B. Mo, J. Mater. Sci. **40**, 6529 (2005)
6. L.F. Kozin, S.N. Nagibin, E.I. Chabanenko, High-Clean Substances (Vysokochistye veshestva) **5**, 30 (1996)
7. L.F. Kozin, S.N. Nagibin, E.I. Chabanenko, A.C. SU 1802533 A1. USSR. 02.01.91.
8. G.S. Rakhymbay, M.K. Nauryzbayev, B.D. Burkitbayeva, A.M. Argimbaeva, A.P. Kurbatov, F. Vacandio, Eurasian Chem.-Technol. J., **16**, 5 (2014)
9. B.D. Burkitbayeva, A.M. Argimbaeva, G.S. Rakhymbay, Kh. Avchukir, G.S. Beisenova, F. Vacandio, D. Tukhmetova, Chem. Bulletin of KazNU, **1**, 7 (2016)
10. S.P. Yatsenko, *Indium Properties and Applications*, (Nauka, Moscow, 1987)

MATEC Web of Conferences **96**, 00006 (2017) DOI: 10.1051/matecconf/20179600006
REE-2016

Numerical Research of Materials Crystal Lattice Parameters Based on Rare-Earth Metals

Artem Obkhodsky[1,*], *Alexander* Popov[1], *Alexey* Zolotarev[2], *Sergey* Kuznetsov[2], and *Victor* Sachkov[3]

[1] Institute of Physics and Technology, Tomsk Polytechnic University, Tomsk, Russian Federation
[2] LLC "Depos", Tomsk, Russian Federation
[3] National Research Tomsk State University, Tomsk, Russian Federation

Abstract. Geometrical parameters (coordinates and angles) of CeO_2 crystal lattice by molecular dynamics method are calculated. Calculated parameters of crystal lattice are applied for definition the energy band structure via Hartree-Fock method in an approximation to CO LCAO (crystal orbitals as linear combination of atomic orbitals) and using the model of cyclic cluster. Calculated minimum energy band p-d is within the value range of experimental data. Valence band maximum is 4.2 while minimum energy band p-d width is 2.8 eV. Quantum-chemical calculations are accelerated by Schwarz inequality and direct inversion method in iterative subspace. The obtained mathematical model is implemented into software package for calculating material properties.

1 Introduction

In view of the fact that the computer capacity is being improved, the calculation of crystal physical properties is becoming a highly topical problem at present. This, in itself, generates interest in the possible reduction of experiments, as such calculations could present the preliminary characteristics of investigated sample and finalize the relevant experimental proof for these characteristics.

In this case such updated methods as Hartree-Fock method and Ab Initio method, molecular dynamics method (MDM) and hybrid methods are applied [1,2]. These methods having different mathematical framework are focused on various research areas. However, it should be noted that Ab Initio method exhibits high calculation accuracy, while MDM- rapid rate.

It is known that about one-third of computer time spent supercomputers to solve problems in the area of atomistic modeling [3]. In this way, currently known mathematical models that describe atomic systems from first principles are resource-intensive. Therein one has the problem of finding new materials - high need of mathematical models in computational resources severely limits the possibility of modeling the atomic structures, in particular - possible number of simulated atoms.

To solve this problem has been developed the software package for materials modeling that supports the methods of molecular dynamics and quantum-chemistry models.

Presently for crystals there are a restricted number of software packages and mathematical models which can be used for description their structures from the first principles. With respect to quantum chemistry, the calculation of molecule reaction properties and electronic crystal properties is more challenging than the calculation of crystal physical properties. In this case, there is a set of mathematical models suitable for posed tasks.

By applying the band theory for solids the following energy bands were defined: maximum and minimum of conduction bands, as well as valence and energy bands. These properties were identified in band diagram (Fig.1).

The aim of this paper is to demonstrate the operability of software package for materials modeling both in terms of molecular dynamics and Ab Initio simulations by the example of CeO_2 crystal. In the present paper geometric coordinates of primitive cells and perfect CeO_2 crystals are examined by molecular dynamics method. The coordinates are calculated on the basis of the energy band structure via Hartree-Fock method in an approximation to CO LCAO (crystal orbitals as linear combination of atomic orbitals) by applying cyclic -cluster model.

2 Mathematical model

One of the most commonly-accepted mathematical models in first-principles calculations of crystal properties is Hartree-Fock method (or Hartree-Fock method in an approximation to CO LCAO - crystal orbitals as linear combination of atomic orbitals).

Regardless of several attempts to accelerate the calculations by Hartree-Fock method [4-6], there are still reasons according to which the calculations of real crystals are difficult. It is a well-known fact that even 1 mm^3 volume crystal contains an immense amount of

* Corresponding author: art707@tpu.ru

atoms, which excludes the possible calculations of real-world samples. This is due to a lack of computing resources for short – time interval calculations [3].

There are several approaches in solving such tasks. For example, a primitive cell is closed by pseudo-atoms on its boundaries and the derived system is calculated. This would be an optimal solution only if the basic criterion involves the shortest calculation time. However, the calculation accuracy, in this case, remains low.

Another approach is based on cyclic-primitive cell and / or cyclic -cluster method. The calculation difficulty of crystals involves the formal infinite number of electrons in the system, while density matrix construction is possible only in case of the finite number of electrons This problem could be solved by imposition of cyclic boundary conditions [7]. It is this approach that is used in the following paper.

Mathematical calculations were taken from famous LCAO-CO method (crystal orbital as a linear combination of atomic orbitals) that is an analogue of the Hartree-Fock-Roothaan method for crystals. On the basis of method results it was possible to connect atom properties which form a crystal with the crystal itself.

Calculation via Hartree-Fock method involves the following sequence:

The first step includes calculating electron kinetic energy:

$$T_{i,j} = -\frac{1}{2}\int \chi_{i,\lambda}(x)\nabla^2\chi_{j,\lambda}(x)dx, \quad (1)$$

further, external potential value:

$$V_{i,j} = \int \chi_{i,\lambda}(x)v(r)\chi_{j,\lambda}(x)dx. \quad (2)$$

Then, the single-electron overlap integrals

$$S_{i,j} = \int \chi_{i,\lambda}(x)\chi_{j,\lambda}(x)dx \quad (3)$$

and electron-electron repulsion integrals are calculated:

$$(ij\,|\,kl) = \iint \frac{\chi_i(r)\chi_j(r)\chi_k(r')\chi_l(r')}{|r-r'|}d^3rd^3r' \quad (4)$$

Hamiltonian operators are given by the following matrix expression:

$$\mathrm{H} = \mathrm{T} + \mathrm{V} \quad (5)$$

Based on the variation principle C^σ is determined from the equation:

$$\sum_j F^\sigma_{i,j}C^\sigma_{j,\lambda} = \sum_j\sum_\mu^{N_\sigma} \delta_{\mu,\lambda}S_{i,j}C^\sigma_{j,\mu}E^\sigma_{\mu,\lambda} \quad (6)$$

Unconstrained Hartree-Fock method is applied including the self-coupling procedure:

1. determining occupied orbital
2. designing density matrix:

$$P^\sigma_{i,j} = \sum_{i,j}^{N_\sigma} n^\sigma_\mu C^\sigma_{i,\mu}C^\sigma_{j,\mu} \quad (7)$$

3. designing Fock matrix:

$$F_{i,j} = T_{i,j} + V_{i,j} + J_{i,j} + K_{i,j} \quad (8)$$

where, $J_{i,j}$ - Coulomb integral:

$$J_{i,j} = \sum_\mu^N \sum_{k,l} nC_{k,\mu}C_{l,\lambda} \times \iint \frac{\chi_{i,\lambda}(x)\chi_{j,\lambda}(x)\chi_{k,\mu}(x')\chi_{l,\mu}(x')}{|r-r'|}d^3xd^3x' \quad (9)$$

$K_{i,j}$ - exchange integral:

$$K_{i,j} = -\sum_\mu^N \sum_{k,l} nC_{k,\mu}C_{l,\lambda} \times \iint \frac{\chi_{i,\lambda}(x)\chi_{l,\lambda}(x)\chi_{k,\mu}(x')\chi_{j,\mu}(x')}{|r-r'|}d^3xd^3x' \quad (10)$$

4. Integrals based on total energy calculation:

$$E = \frac{1}{2}\sum_{i,j}[(P^\alpha_{i,j} + P^\beta_{i,j})H + P^\alpha_{i,j}F^\alpha_{i,j} + P^\beta_{i,j}F^\beta_{i,j}] \quad (11)$$

5. Verification of E – E_{old} < const, where const – constant value, i.e. convergence criterion of self-coupling for each integral of calculated total energy.

LCAO-CO approximation comparable to LCAO MO (old version) that crystal orbitals $\psi^\sigma_i(k,r)$ as Bloch atomic function totals are introduced into the systems with translation symmetry:

$$\Psi^\sigma_i(k,r) = \sum_\mu C^\sigma_{i\mu}(k)\Phi_\mu(k,r) \quad (12)$$

$$\Phi_\mu(k,r) = \sum_n \exp(ikR_n)\varphi_\mu(r - R_A - R_n) \quad (13)$$

In LCAO-CO one-electron functions for the crystal are constructed as linear combinations of atomic orbitals sums in blocks:

$$\varphi_{nk}(r) = \sum_p C_{np}(k)b_{pk}(r). \quad (14)$$

Coefficients $C_{np}(k)$ are found with the help of variational principle from the Hartree-Fock-Roothaan equations:

$$\sum_p F_{pq}(k)C_{np}(k) = E_n(k)\sum_p S_{pq}(k)C_{np}(k), \quad (15)$$

and instead of one such system for a molecule we obtain N systems for crystals.

Matrix $F_{pq}(k)$ is found as a sum of matrices $H_{pq}(k)$ and $G_{pq}(k)$, the elements of which are determined in the following way:

$$H_{pq}(k)=\int b^{*}_{pk}(r)\left[-\frac{1}{2}D-\sum_{A}\frac{Z_A}{|r-R_A|}\right]b_{qk}(r)\ , \quad (16)$$

$$G_{pq}(k)=\sum_{s,t}\sum_{k'}P_{st}(k')[(b_{kp}b_{kq}\,|\,b_{k's}b_{k't})-\frac{1}{2}(b_{kp}b_{k's}\,|\,b_{k't}b_{kq})] \quad (17)$$

Density matrix for crystal is calculated with the formula:

$$P_{st}(k')=2\sum_{m}C_{ms}(k')C^{*}_{mt}(k') \quad (18)$$

Lattice sums of matrix elements calculated with functions from the zero (central cell):

$$F_{pq}(k)=\sum_{n}F_{p0,qn}\exp(ika_n)\,, \quad (19)$$

$$S_{pq}(k)=\sum_{n}S_{p0,qn}\exp(ika_n)\,. \quad (20)$$

While solving the Hartree-Fock-Roothaan equations for crystals it is necessary to know $C_{np}(k)$ of the whole reduced Brillouin zone for fixed value because the specified form of matrix elements as follows:

$$F_{p0,qn}=H_{p0,qn}+G_{p0,qn}\,, \quad (21)$$

$H_{pq}(k)$ – Hamiltonian elements for the crystal which contains energy of electron interaction with cores *V(r)*, and also it contains kinetic energy.

As a result, for self-consistent calculation of the electron crystal structure it is necessary to summarize a large number of occupied electronic states on every stage of integration process.

In the generalized eigenvalue problem the matrices F and S are calculated in the whole reduced Brillouin zone (for fixed value), while Brillouin zone integration in calculating Fock matrix is substituted by summation over spatial points.

Highly precise short-time calculations are one of the major problems in modern quantum chemistry. Applying Hartree-Fock method, including Roothaan equation, (for example, for both MO LCAO and CO LCAO approximations) calculation time is relatively more than in the case if applying density functional theory (proportional N^4, where N – number of basic functions used in calculations). In this case, Roothaan equation is more rarely applied than density functional theory. This problem for Hartree-Fock method is partially solved by direct inversion in iterative subspace and Schwarz inequality application. Thus, total calculation decreases from N^4 to N^2 but calculation precision does not decrease [8].

Schwarz inequality:

$$|(\mu\nu\,|\,\lambda\sigma)|\le\sqrt{(\mu\nu\,|\,\mu\nu)(\lambda\sigma\,|\,\lambda\sigma)} \quad (22)$$

This, in its turn, makes it possible to calculate upper value limits (threshold) for all four two- electron central integrals. This threshold could be ignored in the calculation itself.

In this paper the authors applied DIIS (direct inversion in the iterative subspace) to accelerate self-coupling convergence in Hartree-Fock domain. Its specific feature is plotting on current iteration linear combination of approximated error vectors from the previous iterations (W^{σ}'):

$$W^{\sigma}=F^{\sigma}\cdot P^{\sigma}\cdot S-S\cdot P^{\sigma}\cdot F^{\sigma} \quad (23)$$

$$W^{\sigma\prime}=X^{T}\cdot W^{\sigma}\cdot X, \quad (24)$$

$X=U\cdot s^{-1/2}$, when, $S=U\cdot s\cdot U^{T}$.

Defined coefficients on current iteration are applied for extrapolation of variable functions in the following iteration [9].

Likewise, mathematical manipulations from [7, 10] are used in the calculations, which would provide the calculation of a perfect crystal by applying cyclic-cluster method.

3 Numerical experiment

The first modeling step includes the calculation of CeO_2 crystal geometry by molecular dynamics method.

Molecular dynamics method is applied in automation of input data implementation and in calculations based on Hartree-Fock method. This method defines more precisely the atomic coordinates in the unit cell, which is then used in Hartree-Fock method for further calculations.

Based on molecular dynamics method substance atoms are presented as material points, the interaction of which is described by potential interaction functions. Function types and coefficients are selected for each substance according to experimental thermodynamic data [11].

In molecular dynamics method the time evolution of atom systems is traced by integrating motion equation [12]:

$$F=m\frac{\partial^2 r}{\partial t^2} \quad (25)$$

Solution algorithm basis includes the following:

- coordinates and the velocity of all particles are given at the initial time instant;
- in each successor step new coordinates and particle velocity are calculated for all acting energy.

The following steps are included within the framework of simulating investigated crystalline structure:

- define initial coordinates and atom velocity of modeled system;
- select ensemble;
- determine the heating procedure and select relevant thermostat;
- select simulation time step;

- select used integrating unit;
- select interaction potential.

Interatomic energy for CeO_2 could be the combination of short-acting and short-range energy [13-15]. In our case, Buckingham potential is used to simulate short-range energy in a crystal:

$$\Phi_{ij} = A\exp\left(-\frac{r}{\rho}\right)-\frac{C}{r^6} \quad (26)$$

Coefficients for Buckingham potential is taken from [16].

Long-range interaction energy could be written by Coulomb law:

$$F_{12} = k\frac{z_1 z_2 e^2}{r^2} \quad (27)$$

By applying the above-mentioned potentials a more exact simulation of CeO_2 crystal atom systems could be conducted.

The next MD simulation step is integrating motion equation to determine instantaneous rates and atom position in the system. The most common integrating methods are Verlet integration, accelerated Verlet integration, leapfrog integration and linear multistep method. In our case, Verlet algorithm is used in the integrating motion equation (step equals 1 fs) [17]:

$$r(t+\delta t) = 2r(t) - r(t-\delta t) + \delta t^2 a(t) \quad (28)$$

CeO_2 crystal cluster model (space group Fm3m) was plotted via software package. The number of N atoms in grid block was 1280. Periodic boundary conditions were also applied. Initial atom rate was described by Maxwell distribution at a given temperature.

In simulation the NVE ensemble is used [18]. The system is connected to thermostat Nose-Hoover [19], which, in its turn, keeps constant temperature close to 295K in the system:

$$\frac{dr(t)}{dt} = v(t), \qquad \frac{dv(t)}{dt} = \frac{f(t)}{m} - k(t)\cdot v(t) \quad (29)$$

where modeled temperature 295K; pressure 0 Pa; relaxation of a system – during 1 ns.

After calculating the geometrical parameters of crystal lattices, Hartree-Fock method is used to calculate CeO_2 crystal (transformation group Fm3m) in an approximation to CO LCAO.

4 Results and discussion

Crystal lattice parameters were obtained on the basis of simulation via molecular dynamics method (table 1).

Calculation results showed good agreement with the experimental data (ratio error is 3.29 %), which testify that the method parameters were correct.

Quantum-chemical calculation results are depicted in figure 1 and table 2. Table 2 also includes results of previous studies and experimental data. Calculation method applied in this paper is denoted as CO LCAO CM (crystal orbitals as linear combination of atomic orbitals via cyclic-cluster method).

Table 1. Calculated coordinate results by molecular dynamics method.

Parameter	Experimental structure [20]	Calculation structure
a/b/c (Å)	5.40972 ± 0.00011	5.5879/5.5879/5.5879
α/β/γ (°)	90.0/90.0/90	90.0/90.0/90

Energy band diagram for CeO_2 is illustrated in Fig. 1: X-axis indicates wave vector values in Brillouin zone for several directions, while Y-axis – electron energy state magnitude. Dotted line is Fermi energy. Upper group Ce is 5d orbital, middle Ce 4f orbital and lower O 2p orbital.

Table 2. Calculation results of CeO_2 energy band structure.

Calculation method	Band-gap width p-f (minimum), eV	Band-gap width p-d (minimum), eV	Maximum valence band, eV
CO LCAO CM	2.8	6.5	4.2
PBE [21]	1.5	6.2	3.7
HSE03	3.1 [22]; 3.5 [23]	6.9 [22]; 7 [23]	4.3 [21]
sX-LDA [21]	4/2	6.5	4.4
B3LYP [24]	3.3	7.65	-
PBE0 [21]	3.94	8.08	-
Experiment	3-3.5 [25, 26]	5.5-8 [25, 26, 27]	4.5-5 [25, 26]

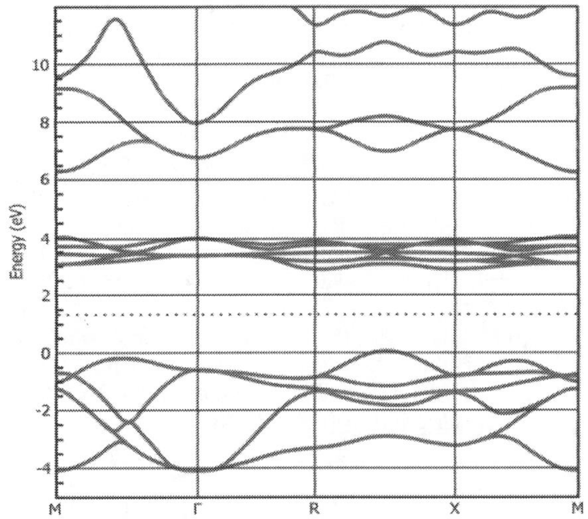

Figure 1. CeO_2 energy band structure.

The calculations are performed over the coordinates obtained by molecular dynamics method, which, in its turn, introduce an error into the calculation itself.

Minimum energy band p-d is 0.2eV less than experimental range limit. Maximum band-gap deviates from the band by 0.3 eV. This could be considered a good result as the coordinates in the calculation are not true. It should be noted that the experiment results are not single numerical values, but a band. This is conditioned by the complexity of the experiment procedure, as the investigated material is fine powder.

5 Conclusion

Operability of software package was confirmed by carrying out numerical experiment. The results of calculation the band structure of CeO_2 ideal crystal were compared with experimentally obtained data for a real crystal, which lattice has defects. Despite this, calculation results are in good agreement with the experimental data. Currently, there are models providing the possible calculation of real crystals [28], which would be further implemented into developed software package for simulation materials properties.

At the present time there is quite a large number of software packages for the simulation of atomic structures. Taking into account the problem, which has been indicated in introduction, the software package will be developed further in the area of scalability calculation resources and optimization the algorithms of mathematical models.

The work was performed with financial support of the Ministry of Education and Science of the Russian Federation. Grant Agreement No. RFMEFI57814X0095, November 28, 2014.

References

1. A.Yu. Kuznetsov, A.B. Sobolev, A.S. Makarov, A.N. Velichko, Solid State Physics (in Russian – Fizika Tverdogo Tela), **11**, 1950 (2005)
2. O.E. Gluhova, I.V. Kirillova, I.N. Salij, A.S. Kolesnikova, E.I. Kossovich, M.M. Slepchenkov, A.N. Savin, D.S. Shmygin, J. Vestnik of Samara State University (in Russian – Vestnik Samarskogo Gosudarstvennogo Universiteta), **9**, 106 (2012)
3. A.Yu. Kuksin, A.V. Lankin, I.V. Morozov, G.E. Norman, N.D. Orekhov, V.V. Pisarev, G.S. Smirnov, S.V. Starikov, V.V. Stegajlov, A.V. Timofeev, Software Systems - Theory and Applications (in Russian – Programmnye Sistemy: Teoriya i Prilozheniya), **1**, 191 (2014)
4. I.A. Obzherin, F.N. Yasinskij, V.V. Sockij, Liquid Crystals and their Use (in Russian – Zhidkie Kristally i ih Prakticheskoe Ispol'zovanie), **4**, 91 (2014)
5. Yu.B. Malyhanov, R.N. Pravosudov, V.V. Meshkov, J. Structural Chem. **41**, 175 (2000)
6. A.I. Svitenkov, D.M. Spel'nikov, V.G. Maslov, A.V. Buhanovskij, Proc. of the Higher Educational Institutions. Instrumentation (in Russian – Izvestiya Vysshih Uchebnyh Zavedenij. Priborostroenie), **12**, 25 (2013)
7. R.A. Evarestov, I.I. Tupicyn, Solid State Phys. (in Russian – Fizika tverdogo tela), **9**, 1582 (2002)
8. D.L. Strout, G.E. Scuseria, J. Chem. Phys, **102**, 8448 (1995)
9. S. Host, J. Olsen, B. Janslik, L. Thogersen, P. Jorgensen, T. Helgaker, The journal of chemical physics, **129**, 124106 (2008)
10. V.A. Veryazov, A.V. Leko, R.A. Evarestov, Solid State Phys. (in Russian – Fizika Tverdogo Tela), **8**, 1407 (1999)
11. J.J. Erpenbeck, W.W. Wood, Modern Theoretical Chem. **6**, 1 (1997)
12. R.M. Kadyshnikov, A.R. Beketov, Powder Met. Ceram. (in Russian – Poroshkovaya Metallyrgiya), **28**, 800 (1989)
13. V.M. Jyrkin, V.I. Poltev, V.L. Florent'ev, Molecular Biology (in Russian – Molecylyarnaya Biologiya), **5**, 1116 (1980)
14. D.W. Brenner, Phys. Rev, **15**, 9458 (1990)
15. W.L. Jorgensen, J. Chem. Phys, **2**, 926 (1983)
16. L. Lambert, Y. Kokou, *Molecular Dynamics Simulations of the Structures of Europium Containing Silicate and Cerium Containing Aluminophosphate Glasses* (University of North Texas, Texas 2012)
17. L. Verlet, Phys. Rev, **159**, 98 (1967)
18. D.J. Evans, G.P. Morriss, Comp. Phys. Rep, **1**, 297 (1984)
19. W.G. Hoover, Phys. Rev, **31**, 1695 (1985)
20. C. Artini, M. Pani, M.M. Carnasciali, M.T. Buscaglia, J.R. Plaisier, G.A. Costa, Inorg. Chem. **8**, 4126 (2015)
21. R. Gillen, S. J. Clark, J. Robertson, Phys. Rev. B, **87**, 125116 (2013)
22. P. Hay, R. L. Martin, J. Uddin, G. Scuseria, J. Chem Phys. **125**, 034712 (2006)
23. J. D. Silva, M. V. Ganduglia-Pirovano, J. Sauer, V. Bayer, G. Kresse, Phys. Rev, **75**, 045121 (2007)
24. J. Kullgren, C. W. M. Castleton, C. Müller, D. M. Ramo, K. Hermansson, J. Chem. Phys, **132**, 054110 (2010)
25. E. Wuilloud, B. Delley, W.-D. Schneider, Y. Baer, Phys. Rev, **53**, 202 (1984)
26. D. Mullins, S. Overbury, D. Huntley, Surface Sci. **409**, 307 (1998)
27. A. Pfau, K. Schierbaum, Surface Sci. **321**, 71 (1994)
28. A.V. Larin, A.N. Kislov, A.E. Nikiforov, S.E. Popov, Solid State Phys. (in Russian – Fizika Tverdogo Tela), **50**, 1378 (2008)

Self-propagating High Temperature Synthesis of Pink Corundum

Anastasiya Nazarova[1], *Nina* Radishevskaya[1], *Nikolay* Kasatsky[1], and *Natalia* Karakchieva[2]

[1]Tomsk Scientific Centre SB RAS, Tomsk, Russian Federation
[2]Tomsk State University, Tomsk, Russian Federation

Abstract. Pigments based on pink corundum have been obtained by self-propagating high temperature synthesis in Cr_2O_3-Al_2O_3-Al system using a small amount of chromium oxide (up to 0.5 wt %) in the composition of the green mixture. Absorption bands of pink corundum in visible reflectance spectrum at λ=560 nm (17850 cm^{-1}) and 400 nm (25000 cm^{-1}) correspond to electron transitions $^4A_{2g}\rightarrow{}^4T_{2g}(^4F)$ and $^4A_{2g}\rightarrow{}^4T_{1g}(^4F)$. Increasing chrome oxide content to 5 wt.% in initial charge mixture leads to pigment darkening due to inclusions of Cr_2O_3. Synthesized pigments can be used as a component of glaze ceramic colors for porcelain.

1 Introduction

Corundum α-Al_2O_3 is a crystalline modification of aluminum oxide of hexagonal symmetry. Corundum lattice represents close hexagonal packing of oxygen ions with insignificant distortions, in which two thirds of octahedral gaps are occupied with aluminum cations. Its structure refers to one of superb compact structure of minerals.

Pure corundum is colorless; however, some coloured forms like rubies and sapphires occur in nature. Coloration is caused by presence of insignificant amount of chromium impurities of d-elements which enable isomorphically replace alumina in its crystalline structure. Pink color of rubies is provided by chromium presence in α-Al_2O_3 (up to 0.5 wt.%). Sapphires occur in a wide variety of colors. Along with blue sapphires there are different colored forms of them. Nickel, titanium, iron (II, III), and manganese impurities yield yellow, blue, orange and reddish pink colors to crystals, correspondingly [1, 2]. All corundum forms have extremely high hardness. According to Mohs scale corundum hardness is 9.

In Table 1 common physicochemical properties of corundum are presented [3].

Table 1. Physicochemical properties of corundum.

Mineral	Ionic radius, Å (Pauling scale)		Density, g/cm^3	Melting temperature, °C	Hardness (Mohs scale)
	Al^{3+}	O^{2-}			
Corundum	0.5	1.4	3.9 ÷4.0	2050	9

Combination of unique properties of corundum such as hardness retained even at temperatures up to 1750°C, excellent dielectric properties, and chemical resistance make it a perfect material for different applications. Corundum is used as a raw material in abrasive industry for fabrication of grinding powder and cutting discs in grinding, polishing and cutting machine applications. It is an essential part of technical ceramics, e.g. insulation elements in electronic devices, ionization chambers, vacuum instruments, fittings of heating elements with operation temperature up to 1750°C. Also clockwork mechanisms are made from corundum.

Transparent forms of colored corundums are used in jewelry production. Colorless synthetic corundum – leucosapphire is utilized in radio electronics. Pink powder of corundum is applied as a pigment in composition of lead-free glaze dyes [4].

Colored corundums in the form of crystals are produced by different ways. Well-known are Verneuil method, Czochralski method, Kyropoulos method, Bridgman–Stockbarger technique, crystallization from vapor phase, crystallization from solution and melts, hydrothermal method, and electric furnace melting [5].

Synthetic rubies are produced in Russia, China, Switzerland, France, USA, Germany, Great Britain, and India. Small amount of sapphires is produced in Japan and Israel.

Synthesis of technical materials is usually conducted by ceramic production method at temperatures 1500-1560°C, required for completing of full transition low-temperature forms of alumina oxides in α-Al_2O_3 with concurrent sintering of particles and overall mass consolidation. Final annealing of high-density products from corundum ceramics is conducted by heating at temperature 1700°C in hydrogen furnace during four hours.

Pink pigment is produced in the furnace at temperature 1500°C by burning of alumina mixture with Cr_2O_3 in the presence of mineralizer (2 wt.% of H_3BO_3) [4].

* Corresponding author: cay@sibmail.com

Self-propagating high temperature synthesis (SHS) is not a traditional way of producing ceramic pigments and refractories based on colored corundums. This method is utilized inner chemical energy of system that provides fast conducting of a chemical reaction and presents oneself an example of power-efficient process [6].

The aim of this work is synthesis of pink corundum by self-propagating high temperature method (SHS) and investigating phase composition and structure of the obtained pigment.

2 Materials and methods

Pigments were synthesized from Cr_2O_3 powder (> 99% pure, "Urals metal", Yekaterinburg, Russia), Al_2O_3 powder (> 99% pure, Soyuzkhimprom, Novosibirsk, Russia) and alumina powder (ASD-4, > 99% pure, LLC Valkom-PM, Volgograd, Russia). Initial components were kneaded thoroughly in a porcelain mortar and after that were placed into the SHS chamber for sintering in flow of argon and oxygen (10% vol.). During combustion the pressure of oxygen was 11.7÷12.7 kPa.

For corundum synthesis porous samples of charge mixture were used. Combustion of powder samples were carried out in furnace on the face end of nichrome electric coil. The heat delivered by electric coil initiated chemical reaction owing to generation of combustion wave. Obtained SHS-derived products were milled.

For producing pink corundum alumina oxide represented by mixture of polymorphous modifications was used. Approximate content of polymorphous modifications of alumina oxide is presented in Table 2.

Table 2. Polymorphous modifications of alumina oxide used for pigment synthesis.

Phase	α-Al_2O_3	γ-Al_2O_3	η-Al_2O_3	θ-Al_2O_3	χ-Al_2O_3
Composition, wt.%	40	11	24	17	8

Calculations have been done with demo version of Match software and PDF-2 and Fityk database application. Table 2 shows that the main part of alumina oxide consists of its low-temperature modifications.

The crystalline phase of obtained pigments were identified by powder X-ray diffraction (DRON-UM, filtered Co K_α-emission) and infrared spectroscopy technique with Fourier transform infrared spectrometer (FTIR) (Thermo Electron Corporation, Nicolet 5700) in the range of 4000-400 cm^{-1}.

Microstructure of SHS-derived products were studied by optical spectroscopy technique with inverted microscope (Carl Zeiss Axiovert 200M). Inorganic pigments based on the pink corundum were subjected to UV-visible spectrophotometry investigation (Thermo Electron Corporation, Evolution 600) equipped with diffuse reflectance accessory.

3 Results and discussions

Pink corundum was synthesized by SHS using the Al_2O_3, Cr_2O_3 and Al powders.

The main reaction in SHS of pink pigment is reaction of alumina oxidation through aluminothermic reaction:

$$Cr_2O_3 + 2Al + O_2 = Al_2O_3 + 2CrO + 444.6\ kJ \quad (1)$$

The initial component Al_2O_3 was used as an inert substance.

Figures 1a-c present microphotographs of pink corundum Cr_2O_3-Al_2O_3. It is seen that absence of chromium inclusions in metallographic sample of pigment indicates that reduction reaction does not occur completely.

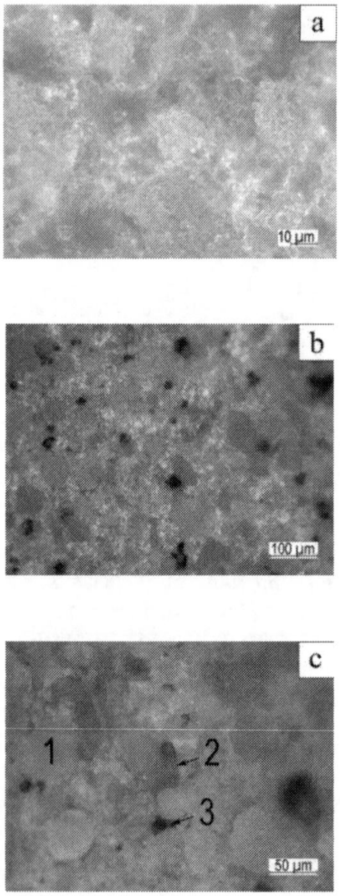

Fig. 1 a, b, c. Microphotographs of pink corundum Cr_2O_3- Al_2O_3 (Axiovert 200M): a) – 0,5 wt.% Cr_2O_3, в) – 5 wt.% Cr_2O_3, c) – 1 wt.% Cr_2O_3, (1) – pink corundum; (2) - large crystal; (3) - Cr_2O_3.

As a result of Cr_2O_3 reduction reaction, the formation of CrO occurs, which is more stable oxide at high temperatures. Binding energy E, calculated for CrO/Cr_2O_3 according to Born-Haber cycle [7], for chromium monoxide is higher than for chromium oxide(III): (E_{Cr2O3}=805 kJ/mol, E_{CrO}=1020 kJ/mol).

Increasing of Cr_2O_3 in initial charge mixture leads to darkening of pigment (Figure 1b). Increasing of Cr_2O_3

DOI: 10.1051/matecconf/20179600007

higher than 5 wt.% causes to occurrence of Cr_2O_3 inclusions in corundum structure after self-propagating high temperature synthesis. As shown in Figures 1 (a, b, c,) corundum has granular structure. Small crystals of red color - ruby - can be found (Figure 1c).

Figure 2 presents XRD patterns of pink pigment. X-ray diffraction analyses showed that there are some reflections which are typical only for corundum.

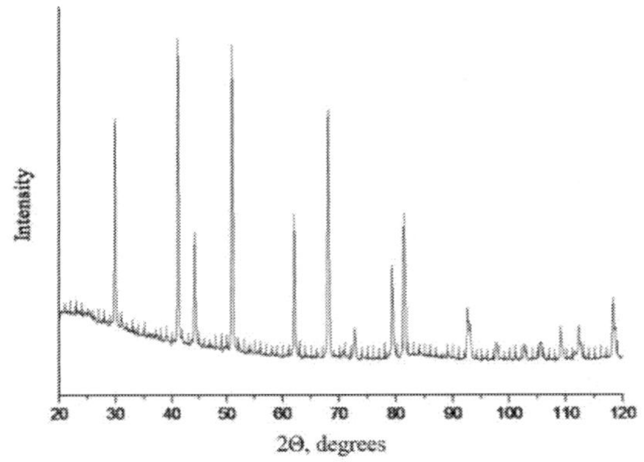

Fig. 2. XRD patterns of pink corundum.

Figure 3 shows reflectance spectra of pink corundum in visible region.

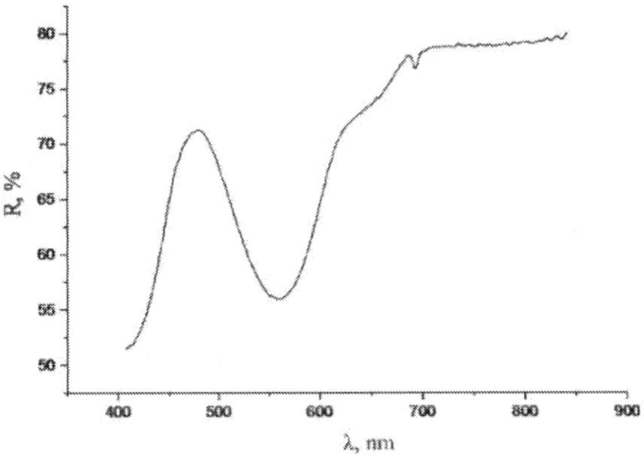

Fig. 3. Reflectance spectra of pink corundum (Cr_2O_3-Al_2O_3 system) in visible region of 40-850 nm.

There are two absorption bands at λ=560 nm (17850 cm^{-1}) and 400 nm (25000 cm^{-1}) in reflectance spectrum of pink pigment that correspond to electron transitions $^4A_{2g}\rightarrow{}^4T_{2g}(^4F)$ and $^4A_{2g}\rightarrow{}^4T_{1g}(^4F)$, and determine intensive pink color of pigment.

Ingress of Cr^{3+} ion impurities into Al-octahedrons that have less size than Cr-octahedrons causes decreasing of chemical bond length between $Cr^{3+}$and O^{2-} due to compression. Ion crystal-chemical radiuses (according to Pauling) for Cr^{3+} and Al^{3+} are 0.64 and 0.5 Å, correspondingly. This leads to increasing of ligand disintegration energy field and transition from green color to red one.

FTIR spectroscopy data analysis according to [9, 10] for obtained pink pigment (0,5 wt.% Cr_2O_3) showed that as a result of SHS, formation of corundum structure occurs which is proved by absorption bands at ~457.4 cm^{-1} , 606.7 cm^{-1} and 643.1 cm^{-1} , typical for corundum phase. It is known that all polymorphic modifications (Al_2O_3) containing in the aluminum oxide powder become α-Al_2O_3 at the temperatures of 1000-1100 °C [8]. The increased content of Cr_2O_3 (0.5 wt %) in the initial reagents leads to a shift of absorption bands up to 476.6 cm^{-1}, 557.5 cm^{-1} and 628.2 cm^{-1}.

However, there are absorption bands for the values of 718.6 cm^{-1} and 1083.6 cm^{-1}, which are typical for tetrahedrally coordinated aluminum $[AlO_4]$ in the spinel $[AlO_4]$ $[CrO_6]$ and in the groups $[AlO_4]$ $[AlO_6]$, indicating that in high-speed SHS processes, not all tetrahedrally coordinated aluminum with low-temperature modifications became octahedrally coordinated (Figure 4).

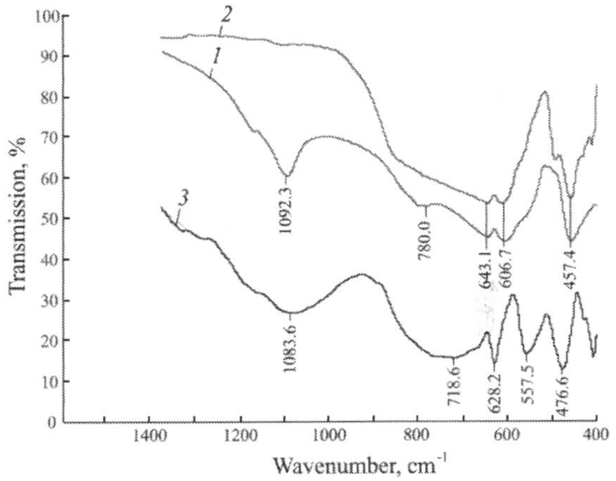

Fig. 4. IR spectra corundum: 1 – grey corundum, 2 - pink corundum (0,5 wt.% Cr_2O_3), 3 - pink corundum (5 wt.% Cr_2O_3).

When the content of Cr_2O_3 oxide is high (≥ 5 wt.% Cr_2O_3) in the green mixture, high temperatures, the reducing behavior and speed of the SHS process can contribute to the formation of the Cr_2+Al_2O_4 spinel, as evidenced by the absorption band $[AlO_6]$ in spinels for 557.5 cm^{-1}. This absorption band can be associated with the aluminum oxide $AlAl_2O_4$ having a spinel structure and described in [11]. The fluctuations of the bonds $[Cr^{2+}O_4]$ overlap on that of the group $[AlO_4]$. Consideration of the available aluminate spinel $AlAl_2O_4$ allows us to assume that the metastable chrome spinel $SrCr_2O_4$ is formed, but it is immediately oxidized to Cr_2O_3 [11]. In the oxygen environment, Cr^{2+} is oxidized to Cr^{3+}. The shift of the absorption band to the high frequency region from 457.4 cm^{-1} to 476.6 cm^{-1} indicates the strengthening of Al-O -bond for the octahedrally coordinated aluminum in corundum due to the substitution of the Al^{3+} cations in the lattice by the larger Cr^{3+} cations. The absorption bands are not observed in the range of ~ 890 - 900 cm^{-1}, indicating that $[Cr^{6+}O_4]$ is not formed [12].

Pigment annealing at temperature 1500°C does not lead to changing of color, which allows using it as a component of glaze ceramic colors for porcelain.

4 Conclusion

Pink pigments were obtained in Cr_2O_3-Al_2O_3 system by self-propagating high temperature synthesis. X-ray diffraction analyses data showed that pigment has corundum structure. Absorption bands of pink corundum in visible region of reflectance spectrum at λ=560 nm (17850 cm^{-1}) and 400 nm (25000 cm^{-1}) correspond to electron transitions $^4A_{2g} \rightarrow ^4T_{2g}(^4F)$ and $^4A_{2g} \rightarrow ^4T_{1g}(^4F)$. Pink color of pigment is caused by ion substitution of Al^{3+} in α-Al_2O_3-octahedron with Cr^{3+} ions. Increasing of chromium oxide in charge mixture higher than 5 wt.% leads to darkening of pigment due to formation of inverted spinel and Cr_2O_3 inclusions. Synthesized pink pigments can be used as a component of glaze ceramic colors for porcelain.

References

1. B.Estifanos, K.Stahl, R.G. Andreassov, G.Bylang, A mineralogical and structural study of red corundum $Al_{1.98}Cr_{0.02}O_3$ from Froland, Norway, Norsk Geologisk Tidsskrift,**77** , 119 (1997)
2. J.M. Duroc-Danner, Untreated yellowish orange sapphire exhibiting its natural color, The Journal of Gemology, **5-8**, 174 (2011)
3. G.V. Samsonov, T.G. Bulankova, F.L. Burykina, T.N. Znatokova, etc. *Physical and chemical properties of oxides. Reference book* (in Russian), (Metallurgy, Moscow, 456, 1969)
4. G.N. Maslennikova, I.V. Pitzh, *Ceramic Pigments* (in Russian), (Stroymaterialy, Moscow, 224, 2009)
5. G.N. Ivanova, *Synthetic analogous of jewelry gems* (in Russian), (Irkutsk University, Irkutsk, 115, 2006)
6. A.S. Rogachev, A.S. Mukasyan, *Combustion for material synthesis: introduction to macrokinetics* (in Russian), (Phyzmatlit, Moscow, 400, 2013)
7. N.P. Lyakishev, Yu. L. Pliner, G.F. Ignatenko, S.I. Lappo, *Aluminothermy* (in Russian), (Metallurgy, Moscow, 424, 1978)
8. A.B. Nikol'sky, *Physical method of inorganic materials investigation* (in Russian), (Akdemiya, Moscow, 448, 2006)
9. R.Yu. Zinyuk, A.G. Balykov, I.B. Gavrilenko, A.M. Shevyakov, *FTIR-spectroscopy in inorganic technology* (in Russian), (Khimiya, Leningrad, 160, 1983)
10. K.V. Chernyakova, I.A. Vrublevskii, M.I. Ivanovskaya, D.A. Kotikov, Impurity defective structure of anodic alumina oxide produced by method of two-sided anodic oxidation in solution of tartaric acid, Journal of Applied Spectroscopy (in Russian), **1**, 83 (2012)
11. Horoshavin L.B. *Spinel Firebriks*. (Ekaterinburg, Ural Branch RAS, 600, 2009).
12. Barabanov V.F. *Modern Physical Methods in Geochemistry*. (Leningrad State University, 391, 1990)

MATEC Web of Conferences **96**, 00008 (2017) DOI: 10.1051/matecconf/20179600008
REE-2016

Periodic Sorption of Tungstate Ions on Anionite AV-17-8

Aleksandr D'yachenko, *Roman* Kraidenko, *Yury* Perederin*, *Mariya* Suprunenko, and *Sergey* Chegrintsev

National Research Tomsk Polytechnic University, Tomsk, Russian Federation

Abstract. The multiple sorption of sodium tungstate resulting from the autoclave-soda digestion of a tungsten-bearing concentrate was studied using anion-exchange resin AV-17-8. The choice of ion exchange resin was carried out under static conditions using highly basic anionites. The sorption and desorption plots for tungstate and carbonate ions were demonstrated under dynamic conditions. The total dynamic capacity of the resin was estimated for each species of the ions in three sorption cycles. The applicability of the AV-17-8 resin as a sorbent in the autoclave-soda process flowsheet was determined.

1 Introduction

Tungsten is applied in illumination devices, cathode ray tubes, as electrodes, and for the manufacture of high-strength and inert alloys, for example, tungsten carbide-based alloys and tungsten-doped pseudo-alloys, because it is refractory and ductile [1-2]. Tungsten is an essential construction material for the thermonuclear experimental reactor (ITER) [3].Tungsten is also used in mechanical engineering, chemical industry, aviation, medicine, and electrochemical processing [4-5].

The usage of tungsten is growing year after year, and investments are increasingly being made into new tungsten mining projects around the world [6]. However, the tungsten production in Russia does not expand proportionally to the tungsten consumption [7].

The tungsten price per 1 kg rose from 10$ (2000) to 56$ (2014), which is indicative of the grown demand for tungsten and of insufficient offers [8]. The global tungsten production volumes account for about 25 000 tons a year [9].

The primary sources of tungsten are wolframite and scheelite [10]. However, tailings of tungsten-bearing sands are of value as well because improved technologies for beneficiation of low-grade tungsten ores can augment the tungsten output up to a saleable tungsten concentrate [11]. Tungsten can also be recovered from wastes of hard-alloy materials [12].

A lot of tungsten production techniques are currently available. The autoclave-soda digestion [13] and acid leaching [14] are considered to be classical techniques for the processing of tungsten-containing ores. New processing techniques for tungsten-bearing concentrates include the roasting of tungsten-bearing concentrates with silicon oxide in the flowing steam-air mixture to transform them into tungsten trioxide [15]; fluorination process by ammonium bifluoride [16]; and exotic methods for recovering tungsten trioxide, for example, solar production of tungsten trioxide [17].

At this point, among the most common digestion techniques for the concentrate is the autoclave-soda process. One of the stages involved in this process is the ion-exchange extraction of tungstate and carbonate ions [18]. The technology utilizes highly basic anionites in the ion exchange [19].

However, there exist cation-exchange methods for the separation of sodium ions from sodium tungstate [20]. Different instrumentation can also be used in the ion-exchange process: classical passage of the solution through ionite under the influence of gravity and ascendant passage of the solution through ionite under the influence of extraneous forces [21].

Anionites AN-106TP and AN-1TP are of use as the ion-exchange resin for the separation of tungsten and sulfate ions [22]. Trace quantities of niobium, tantalum, tungsten, zirconium, and hafnium can be separated on anionite Dowex 1X8-100 [23].

The main problem of the ion-exchange extraction of tungsten is that ionite must endure as many sorption-desorption cycles as possible, but in fact the secondary sorption may take place incompletely. Hence, necessity exists to investigate the sorption conditions of tungstate and carbonate ions from solutions after autoclave-soda digestion.

The aim of this work is the tungstate ions selection from the tungsten solution after autoclave-processed soda for further obtaining the tungsten oxide and soda regeneration for its re-use as a leaching agent.

Two tasks have been solved with this goal:

- choosing a suitable anion exchanger;
- checking the anion exchanger under the conditions of repeated sorption.

2 Experimental part

Different ion-exchangers were comparatively evaluated for the capability of tungstate and carbonate ions to be sorbed and desorbed from the solution after autoclave-soda digestion.

* Corresponding author: perederinyv@rambler.ru

The composition of the clarified, concentrated tungsten-containing solution after autoclave-soda digestion is given in Table 1. Tungsten-containing solution was prepared by autoclave-processed soda.

Table 1. Composition of tungsten-containing solution after autoclave digestion.

Composition	**Ca**	**K**	**Na**	**Si**	**W**	**CO_3^{2-}**
Concentration, g/L	0,07	0,16	63,65	0,13	42,95	75,50

The following highly basic anionites were used as ion-exchangers: Purolite A-400, A-500, A-600 and Ambersep 920-U, AV-17-8. The anionites were saturated under dynamic conditions with the concentrated tungsten-containing solution in a solid-to-liquid ratio of 1:1 for 1 h. Desorption was performed with a 2 M ammonium chloride solution in the same solid-to-liquid ratio of 1:1 for 1 h.

The tungsten concentration was measured by atomic emission spectrometry and the carbonate-ion concentration by titration [24].

Ambersep 920-U exhibited the highest sorption of tungstate and carbonate ions, but desorption on this desorbent was much worse than on the other resins. Specifications of the used anion-exchange resins are listed in Table 2.

Table 2. Specifications of anion-exchangers.

No	**Name**	**Appearance**	**Ionic form**	**Total exchange capacity, g eq/dm^3**	**Average size, mm**	**Uniformity factor**
1	Purolite A-400	Non-transparent spherical particles	Cl-	0,8	0,3 – 1,2	1,6
2	Purolite A-500	Non-transparent spherical particles	Cl-	0,8	0,3 – 1,2	1,6
3	Purolite A-600	Non-transparent spherical particles	Cl-	1,4	0,3 – 1,2	1,7
4	Ambersep 920-U	Transparent spherical particles	Cl-	1,0	0,75 – 0,95	1,5
5	AV-17-8	Yellow spherical particles	Cl-	1,15	0,31 – 1,25	1,7

3 Results and discussion

The amounts of sorbed and desorbed tungstate ions on different anionites are presented on Figure 1.

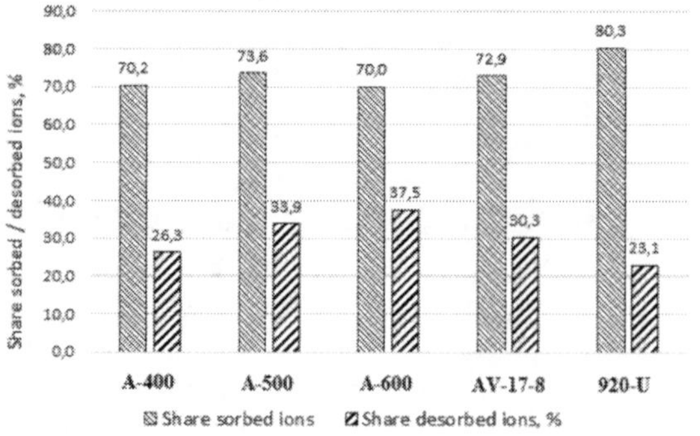

Fig. 1. Distribution of sorbed and desorbed tungsten ions on different anionites.

The percentages of sorbed and desorbed carbonate ions on different anionites are depicted in Figure 2.

The sorption of carbonate and tungsten ions from the clarified concentrated solution was examined after autoclaving with soda digestion.

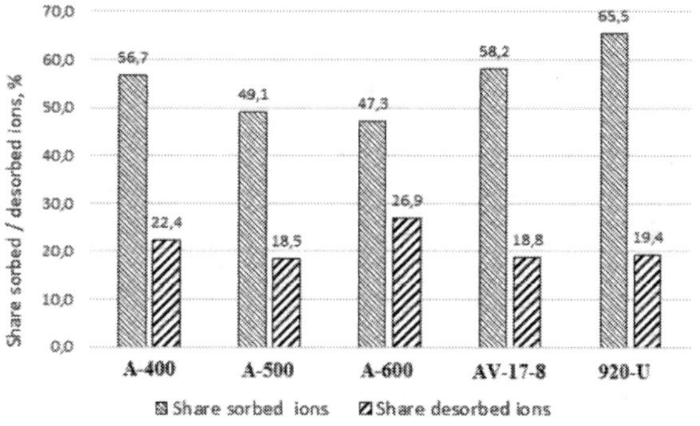

Fig. 2. Distribution of sorbed and desorbed carbonate ions on different anionites.

AV-17-8 anionite comprising the benzyltrimethylammonium ionogenic group was employed as the ion-exchange resin. This anionite was chosen as a high-grade reagent that has high static and dynamic exchange capacities.

The anionite was rendered into the chloride form by reaction with a 10% solution of ammonium chloride.

A total of three sorption-desorption cycles were performed. In each experiment, the tungsten-containing solution volume was 100 mL, the ionite weight was 15 g, and the ionite volume was 25 mL. The propagation velocity of the solution through the ionite was 1 drop per 3 sec, and the propagation of the solution through one column volume was 45 min. A 2 mol/L ammonium chloride solution (100 mL) was used as the desorbent.

Forty percent of tungstate ions are sorbed from the concentrated sodium tungstate solution in the 1st cycle, 38% in the 2nd cycle, and 32% in the 3^{d} cycle. Due to the high concentration, the tungstate ion is unable to be sorbed fully on the resin under the specified conditions. The decreased velocity makes the sorption process limiting; therefore, the ion exchange should be performed in dilute solutions.

DOI: 10.1051/matecconf/20179600008

The difference in tungstate ions sorption in the three cycles is negligible (Figure 3).

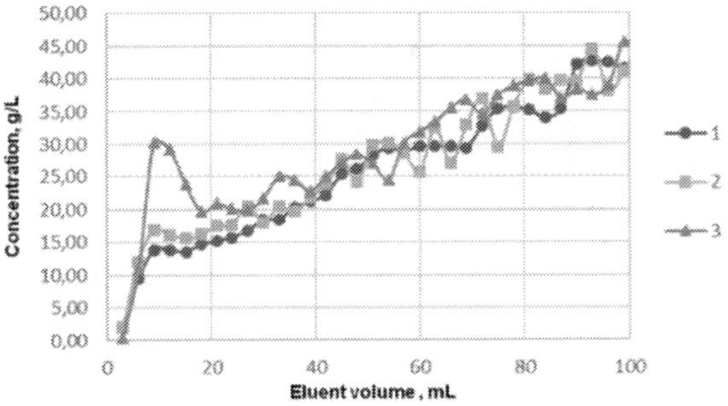

Fig. 3. Tungsten concentration in the eluate after sorption plotted against eluent volume for three cycles; 1 - in the 1st cycle, 2 - in the 2nd cycle, 3 - in the 3d cycle.

The dynamic capacity of the anionite against tungsten diminishes: 320 g eq/L for the first sorption cycle, 278 g eq/L for the second, and 237 g eq/L for the third. This phenomenon is explained by incomplete desorption. Desorption against tungsten is 100% in the first cycle and only 70% in the second. The third cycle shows again 100% desorption against tungsten, but because the resin capacity decreases with each cycle, carbonate ions fill the space of tungstate ions in the resin matrix.

Desorption of tungstate ions is considerably distinct in each of the cycles. Desorption is uniform in the first cycle, whereas the ions are desorbed chiefly with first portions of the eluent in the second and third cycles (Fig. 4).

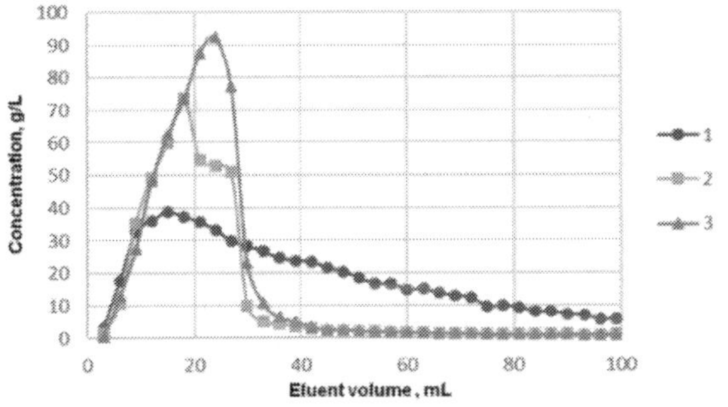

Fig. 4. Tungsten concentration in the eluate after desorption plotted against eluent volume for three cycles; 1 - in the 1st cycle, 2 - in the 2nd cycle, 3 - in the 3d cycle.

The sorption of carbonate ions differs from that of tungstate ions. For instance, the sorption capacity of the resin against the carbonate ion declines in the third cycle, and the resin does not nearly absorb carbonate ions (Fig. 5).

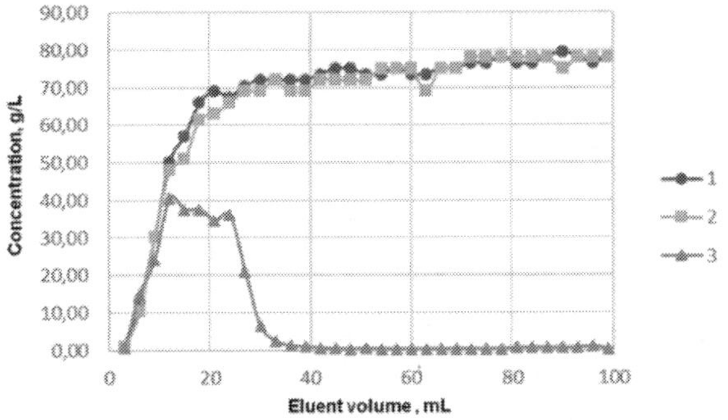

Fig. 5. Carbonate ion concentration in the eluate after sorption plotted against eluent volume for three cycles; 1 - in the 1st cycle, 2 - in the 2nd cycle, 3 - in the 3d cycle.

Desorption profile for carbonate ions follows that of tungstate ions (Fig. 6). Desorption is observed to be more uniform in the first cycle, while it is jump-like in the third cycle.

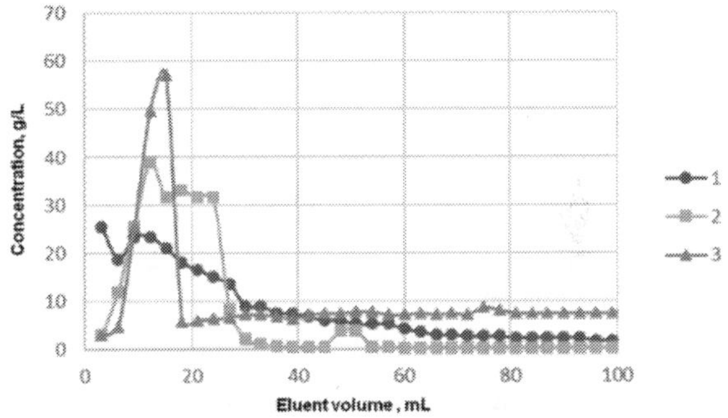

Fig. 6. Carbonate ion concentration in the eluate after desorption plotted against eluent volume for three cycles; 1 - in the 1st cycle, 2 - in the 2nd cycle, 3 - in the 3d cycle.

The anionite capacities against the carbonate ion in the first two cycles do not differ and are 184 g eq/L; the capacity in the third cycle is only 73 g eq/L of the initial. The carbonate ions are presumably converted into hydrocarbonate ions which cannot be desorbed with ammonium chloride. In turn, the hydrocarbonate ions take up the place of the mobile ions in the resin matrix and do not allow the sorption of both tungstate ions and carbonate ions.

4 Conclusion

It can be concluded from the findings that the sorption of the tungstate ions from the solution on the AV-17-8 resin in the first sorption cycle after autoclave-soda digestion is 40%; however, the anionite capacity then diminishes due to incomplete desorption of the carbonate ions that are also presented in the tungsten-containing solution.

The AV-17-8 resin is not fit for multiple sorption. For effective use of the AV-17-8 resin and effective sorption of all the tungstate ions, the concentration of the target component in the initial solution should be decreased by water dilution.

This research was conducted with support from the Russian Federation Government (Russian Ministry of Education and Science), contract No. 02.G25.31.0118.

References

1. V.P. Reva, D.V. Onishchenko, V.A. Kim, V.V. Petrov, A.I. Evstigneev, Refract. Ind. Ceram+ **54**, 295 (2013)
2. J. Wilden, V.E. Drescher, M. Schütze, R. Durham, Materialwissenschaft und Werkstofftechnik **41**, 608 (2010)
3. J. Matějíček, B. Nevrlá, M. Vilémová, H. Boldyryeva, Nukleonika **60**, 267 (2015).
4. A.N. Zelikman, *Metallurgiya Vol'frama i Molibdena [Metallurgy of Tungsten and Molybdenum]* (Metallurgizdat, Moscow, 1949) (in Russian)
5. N. Holstein, W. Krauss, J. Konys, Fusion Eng. Des. **86**, 1611 (2011)
6. A.S. Sánchez, A. Krzemień, P.R. Fernández, F.J. Iglesias Rodríguez, F.S. Lasheras, F. Javierde Cos Juez, Resources Policy **46**, 177 (2015)
7. S.S. Bel'skiy, O.E. Bel'skaya, Vol'fram – Proizvodstvo i Potrebleniye [Tungsten – Production and Consumption] *in Proc. of 3^{d} Inter. Science and Technology Conference "Metallurgy of Light and Refractory Metals"* (Ural Federal University, Yekaterinburg, 2014) (in Russian)
8. N.I. Novikov, V.A. Salikhov, Vestnik Tomskogo Gosudarstvennogo Universiteta. Ekonomika **2**, 138 (2015) (in Russian)
9. I.A. Rodionova, *Makrogeographiya Promyshlennosty Mira: Uchebnik [Macrogeography of the World Industry: Textbook]* (Moskovskiy Litsei, Moscow, 2000) (in Russian)
10. C.J. Smithells, *Tungsten: A Treatise on Its Metallurgy, Properties and Applications*, 3rd ed. (Chapman and Hall Ltd., London, 1952)
11. A.N. D'yachenko, S.I. Ivankov, R.I. Kraidenko, A.B. Manucharyants, D.G. Petkevich-Sochnov, G.S. Spitsin, Yu.V. Perederin, A.G. Karpov, V.Yu. Yegorov, Polzunovskiy Vestnik **2**, 120 (2015) (in Russian)
12. A.I. Nikolaev, V.G. Maiorov, V.K. Kopkov, Russ. J. Appl. Chem+ **70**, 521 (1997)
13. A.N. Zelikman, G.A. Meerson, *Metallurgiya Redkikh Metallov [Metallurgy of Rare Metals]*, (Metallurgiya, Moscow, 1973) (in Russian)
14. A.S. Medvedev, *Vyshchelachivaniye i Sposoby Ego Intensifikatsii [Leaching and Its Intensification Methods]* (MISIS, Moscow, 2005) (in Russian)
15. S.S. Bel'skiy, Vestnik Irkutskogo Gosudarstvennogo Tekhnicheskogo Universiteta **12**, 204 (2015) (in Russian)
16. E.I. Kurchenko, Chemical Concentration of Low-grade Tungsten-containing Industrial Waste Discharge *in Proc. of Inter. Symposium "Inorganic Fluorides: Chemistry and Technology"* (TPU Publishing House, Tomsk, 2014)
17. H.I.V. Vidales, A. Jiménez-González, Green Process. Synth. **4**, 167 (2015)
18. A.N. D'yachenko, A.P. Dugel'nyi, R.I. Kraidenko, S.N. Chegrintsev, Izvestiya Tomskogo Politekhnicheskogo Universiteta **322**, 62 (2013) (in Russian)
19. R.I. Kraidenko, Yu.V. Perederin, D.S. Filatov, A.B. Manucharyants, A.G. Karpov, M.S. Vasilishin, Polzunovskiy Vestnik **2**, 135 (2015) (in Russian)
20. Y.G. Choi, G. Sakai, K. Shimanoe, N. Miura, N. Yamazoe, Sensor Actuat B-Chem. **87**, 63 (2002)
21. V.S. Yu, S. Nakai, A. Ali, The analyst **131**, 434 (2006)
22. A.G. Kholmogorov, V.P. Kyrillova, O.N. Kononova, S.V. Kachin, O.P. Kalyakina, G.L. Pashkov, Hydrometallurgy **53**, 177 (1999)
23. A.G. Coedo, T.D. Lopez, F. Alguacil, Anal. Chim. Acta **315**, 331 (1995)
24. GOST (Russian State Standard) Nos. 20301-74, 13504-68, 5.1342-72: Smoly Ionobmennyye. Anionity [Ion Exchange Resins. Anionites] (Izdatel'stvo Standartov, Moscow, 1992)

Production of Nd-Fe Foundry Alloy by Electrolysis in Molten Salts

Yuriy Makaseev[1], *Alexander* Buinovskiy[1,2], *Sergey* Zhitkov[1], *Evgeniy* Kartashov[1], and *Vladimir* Sofronov[1,*]

[1]National Research Nuclear University "Moscow Engineering Physics Institute", Moscow, Russian Federation
[2] National Research Tomsk State University, Tomsk, Russian Federation

Abstract. Nd-Fe-B-based permanent magnets have the highest magnetic characteristics. To improve their quality, one generally applies the technique of solid-phase alloying by materials with high content of neodymium, dysprosium, terbium, i.e. rare-earth metal (REM)-Fe(Co) foundry alloys. The paper presents the results of the study of the manufacturing process of Nd-Fe foundry alloys, with composition being close to eutectic, by electrolysis of neodymium oxides and fluorides from chloride and fluoride melts. It is shown that in the oxide electrolysis a sufficiently high current efficiency (over 98 %) is obtained. This is a promising method of obtaining of rare earth-containing materials.

1 Introduction

In Russia there has been developed the fluoride method of production of Nd-Fe-B-based magnetic alloys and rare earth foundry alloys (Nd, Pr, Dy, Tb), with a transition metal being Fe or Co. The method is based on the out-of-furnace calcium-thermic coreduction of anhydrous metal fluorides. The advantages and disadvantages of this method are presented in the paper [1].

Nd-Fe-B alloys are used for the production of high-energy permanent magnets. Foundry alloys are used either for the production of magnetic alloys by refining or for solid-phase alloying of magnetic alloys during the stage of their comminution, when magnets are produced by means of the powder metallurgy method.

The fluoride technique allows obtaining foundry alloys in the form of ingots; during the reduction smelting the yield is about 95% when iron and neodymium contents in the foundry alloy are close to calculated ones for the formation of low-melting eutectic (76.5% wt. Nd-23.5Fe with the melting temperature $T_m \approx 640$-650°C) [2]. However, the fluoride technique has some disadvantages:

• the yield of foundry alloy is not high enough during the reduction smelting;
• the formation of fluoride slags (about 1 kg per 1 kg of the ligature); they are generally composed of CaF_2 and it is hard to process them;
• considerable material and labour costs, connected with the production of anhydrous metal fluorides, calcium metal, graphite crucible manufacturing for the reduction smelting and the complication of the continuous process organization during main production steps.

In this regard, the search for better and economically sound methods of production of magnetic alloys, including Nd-Fe foundry alloy, is an urgent task for the further development of high-energy magnet production.

Therefore, along with the improvement of the out-of-furnace fluoride technology we develop the electroplating technique for producing REMs (rare-earth metals) and foundry alloys, such as fusible Nd-Fe foundry alloy. Electroplating techniques of REM and foundry alloy production are based on the electrolysis of fluorides, oxides or the mixture of fluorides and chlorides of alkali and earth metals in molten salts. At the same time, a carbon anode and an iron cathode are generally used when the electrolysis of neodymium fluoride or neodymium oxide is taking place [3-7]. The process is conducted at 650-1100 °C.

In our opinion, the electroplating technique of foundry alloy production has some advantages over the fluoride technology:

• the electrolysis process is controllable, and it allows controlling the progress of the process with the help of inspection tools and automatic equipment;
• neodymium fluoride consumption decreases; and application of an iron cathode, while using neodymium oxide as consumable supplies, excludes neodymium fluoride consumption;
• there is a real possibility of creating an electrolyzer for continuous operation with dosed charge of consumable supplies (neodymium oxide and neodymium fluoride) and periodic discharge of the molten foundry alloy during electrolysis; that will dramatically reduce the amount of wastes in the form of spent electrolyte.

2 Experimental part

The research was conducted in the experimental electrolyzer, which principle circuit is presented in the Fig. 1. The reaction crucible had the form of a truncated cone with a hermetic cover made of metal tungsten.

* Corresponding author: VLSofronov@mephi.ru

 DOI: 10.1051/matecconf/20179600009

The following electrodes were put into the crucible of the electrolyzer: two graphite anodes and a cathode that can be made of different materials (a tungsten cathode is for the production of metal neodymium; an iron cathode is for the production of Nd-Fe foundry alloy). Ready electrolyte, consisting of metal fluorides or chlorides and neodymium oxide, was charged into the crucible. The electrolyzer was placed into a resistance furnace with the operating heating temperature of 1300 °C.

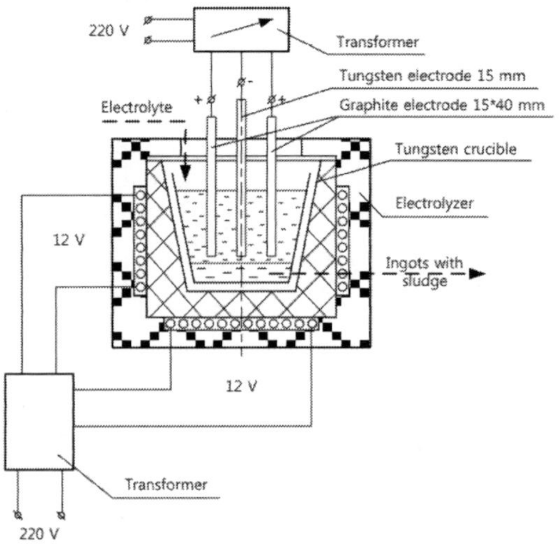

Fig. 1. Principle circuit of the electrolyzer.

The electrolyzer was assembled and charged with salts; after the preliminary vacuumization it was filled with argon; then it was heated up to the salt melting temperature and direct current flowed through the molten salts. During electrolysis the electrolyte was stirred by an agitator. To neutralize the gases, evolved during the electrolysis at the anode, they were run through the barium hydroxide solution.

The ingot of Nd-Fe foundry alloy, obtained by electrolysis, was tested for the content of neodymium, iron and other elements.

During the electrolytic production of the foundry alloy, neodymium fluoride and neodymium oxide were used as consumable compounds of neodymium. Their chemical composition is presented in the table 1.

Table 1. Chemical composition of neodymium fluoride and neodymium oxide.

Neodymium compounds	Chemical composition of neodymium fluoride and neodymium oxide, wt%			
	Nd	Pr	Cu	Ni
Nd_2O_3	77.7	1.3	0.06	0.03
NdF_3	69.6	1.2	0.03	0.3
	Ca	Al	Mg	F
Nd_2O_3	0.009	0.009	0.3	-
NdF_3	0.06	0.02	0.3	27.9

2.1 Production of Nd-Fe foundry alloy by the electrolysis of neodymium fluoride

Fluoride (NdF_3 – LiF – BaF_2) and chloride-fluoride ($BaCl_2$ – LiCl – NdF_3 – LiF) molten salts, containing 25-40 wt% NdF_3, were tested as an electrolyte for the electrolysis of neodymium fluoride. During the electrolysis of these melts at the electrodes the following reactions are taking place:

at the graphite anode:

$$2F^{-1} - 2e \rightarrow F_2 \text{ (gas)},$$
$$C + 2F_2 \rightarrow CF_4 \text{ (gas)},$$
$$2Cl^{-1} - 2e \rightarrow Cl_2 \text{ (gas)},$$

at the iron cathode:

$$Nd^{+3} + 3e \rightarrow Nd,$$
$$Nd + Fe \rightarrow Nd_2Fe + \alpha\text{-}Nd \text{ – low-melting eutectic}$$
$$(T_m = 640°C).$$

Figures 2 and 3 show the influence of the cathodic current density ($i_к$) on the current efficiency ($\eta_т$) for neodymium into the cathodic product during NdF_3 electrolysis.

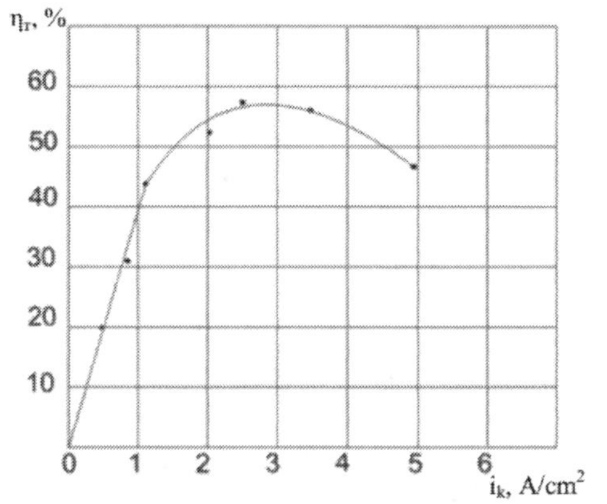

Fig. 2. Dependence of the current efficiency for neodymium on the cathodic current density at the temperature of the process t = 850°C (electrolyte NdF_3-LiF-BaF_2).

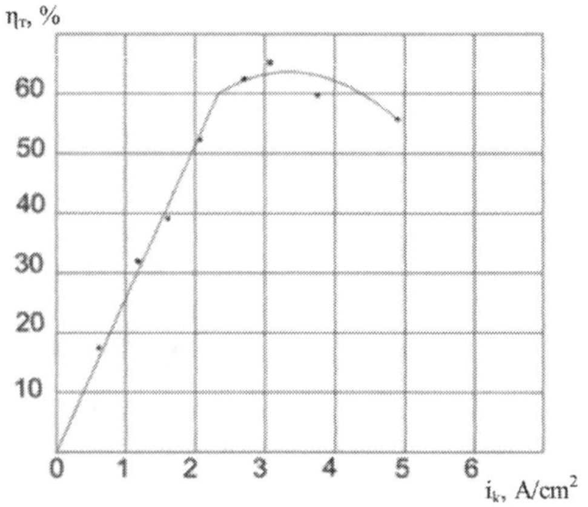

Fig. 3. Dependence of the current efficiency for neodymium on the cathodic current density at the temperature of the process t = 750°C (electrolyte $BaCl_2$-LiCl-NdF_3-LiF).

During the electrolysis of NdF_3 in fluoride and chloride molten salts, the neodymium content in the cathodic product was changing from 78.2 to 80.1wt%. At the same time the optimal current efficiency for neodymium was $\eta_т \approx 58\text{-}60$ % during electrolysis in the fluoride electrolyte and $\eta_т \approx 62\text{-}66$ % - in the chloride-fluoride electrolyte.

2.2 Production of Nd-Fe foundry alloy by the electrolysis of neodymium oxide

As an electrolyte for the electrolysis of neodymium oxide there have been tested fluoride molten salts of $NdF_3 - LiF - BaF_2$ composition, containing 40-60 wt% NdF_3, in which solubility of Nd_2O_3 reached 5-7 wt%, which is sufficient for the oxide electrolysis. During the electrolysis of neodymium oxide in fluoride melts at the electrodes the following reactions are taking place:

at the graphite anode:

$$2O^{-2} - 4e \rightarrow O_2,$$

$$C + O_2 \rightarrow CO_2\ (gas),$$

$$2C + O_2 \rightarrow 2CO\ (gas),$$

at the iron cathode:

$$Nd^{+3} + 3e \rightarrow Nd,$$

$$Nd + Fe \rightarrow Nd_2Fe + \alpha\text{-}Nd - \text{low-melting eutectic.}$$

Figure 4 shows the influence of $i_к$ on $\eta_т$ during the electrolysis of neodymium oxide in the fluoride electrolyte.

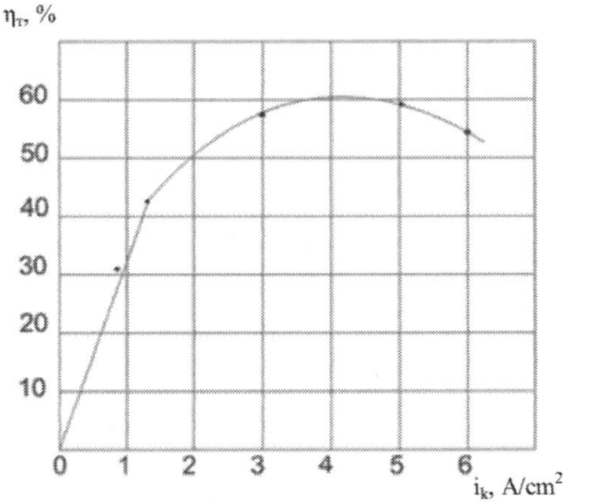

Fig. 4. Dependence of the current efficiency for neodymium on the cathodic current density at the temperature of the process t = 850°C (electrolyte BaF_2-NdF_3-LiF).

The electrolysis of neodymium oxide was stopped after precipitation of the sparingly soluble barium carbonate from the solution-neutralizer due to the reaction taking place between carbon dioxide, evolving at the graphite anode, and the barium hydroxide solution:

$$CO_2 + Ba(OH)_2 \rightarrow BaCO_3 + H_2O.$$

The optimal current efficiency for neodymium was $\eta_т \approx 58\text{-}60$ % during Nd_2O_3 electrolysis in the fluoride electrolyte.

Figure 5 shows the history of the electrolysis of neodymium oxide in time. Voltage variation at the iron cathode during NdF_3 electrolysis with periodicity of 5-7 minutes has stick-slip nature. This is due to detachment of droplets of liquid Nd-Fe foundry alloy from the cathode to the bottom of the electrolysis cell (the tungsten crucible).

Table 2 presents the results of experiments on neodymium oxide electrolysis in the fluoride electrolyte BaF_2–NdF_3–LiF in the selected mode with neodymium oxide loading during electrolysis.

Conditions for electrolysis:

- the electrolyte composition, wt%: $13BaF_2 - 60NdF_3 - 27LiF$,
- parameters: $i_A = 0.8\text{-}1.0$ A/cm^2, t = 850°C, τ = 6 hours,
- portions of Nd_2O_3, weighing 15-20 g, were immersed into the electrolyte every 60 minutes.

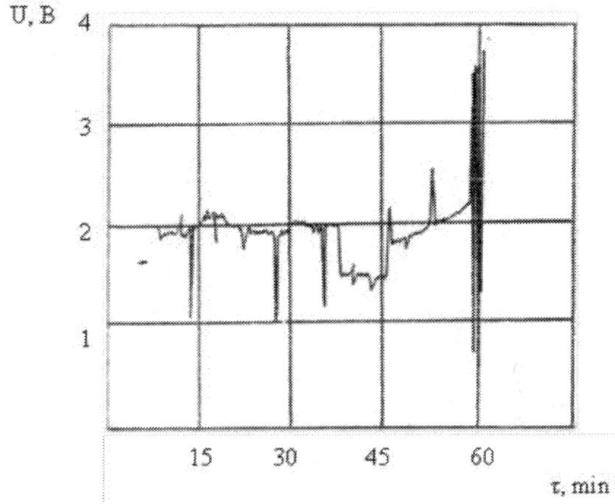

Fig. 5. Potential variation of the cathode in time.

Table 2. Results of experiments on neodymium oxide electrolysis in the fluoride electrolyte.

No	Weight of given electrolyte, g	Processed, g: Nd_2O_3	Processed, g: Nd with Nd_2O_3	Obtained Nd-Fe foundry alloy, g
1	620	100	73.8	90.4
2	600	150	110.7	128.3
3	700	100	73.8	82.3
4	700	150	110.7	130.6
5	650	130	95.9	110.0
6	600	100	73.8	90.8

No	Foundry alloy composition, wt%: Nd	Foundry alloy composition, wt%: Fe	Amount of Nd in Nd-Fe foundry alloy, g	Current efficiency, $\eta_т$, %
1	80.3	14.3	72.6	98.4
2	83.3	13.9	107.0	96.7
3	82.4	13.8	67.8	91.9
4	83.7	16.8	109.3	98.7
5	81.4	15.2	89.6	93.4
6	80.3	16.1	72.9	98.8

Table 3 presents the results of the analysis of the Nd-Fe foundry alloy ingot, obtained by neodymium oxide electrolysis in the molten fluoride salts (BaF_2–NdF_3–LiF).

DOI: 10.1051/matecconf/20179600009

Table 3. Chemical composition of the Nd-Fe foundry alloy ingot.

Element	Nd	Fe	Pr	Cu	Ni	Al	C
Content, wt%	83.4	14.0	2.2	0.15	0.15	0.03	0.02

Data presented in table 3 show that the overall content of all impurities in the foundry alloy does not exceed 0.4 % wt.

3 Conclusion

The study of the electroplating technique of production of Nd-Fe foundry alloy in fluoride, chloride-fluoride and oxide-fluoride molten salts has been carried out;

• it has been determined that during electrolysis of Nd_2O_3 and NdF_3 in BaF_2–NdF_3–LiF melt and during electrolysis of NdF_3 in $BaCl_2$–LiCl–NdF_3–LiF melt, low-melting eutectic of iron and neodymium is formed at the iron cathode. The content of the low-melting eutectic of iron and neodymium is close to the calculated composition of the foundry alloy;

• under optimum conditions of Nd_2O_3 electrolysis in a fluoride molten salt, neodymium yield into the cathode product reaches 98.4-98.8%, which is significantly higher than during NdF_3 electrolysis; thus, this method is very promising for production of materials containing Nd and, apparently, other rare-earth metals.

The work was made under the aegis of the Federal goal-oriented program "Researches and design of foreground directions of development of Russia's scientific-and-technological system in 2014-2020" (RFMEFI57814X0018).

References

1. A. Bujnovskij, V. Sachkov, V. Sofronov, A. Anufrieva, Adv. Mater. Res. J. **1085**, 209 (2015)
2. F.J.G. Landgraf, G.S. Schneider, V. Villas-Boas, F.P. Missel, Less-Common Met. J. **163**, 209 (1990)
3. V. Grebnev, V. Dmitrienko, Bul. of the TPU J. **311**, 70 (2007)
4. V. Soare, M. Burada, T. Ostvold, C. Kontoyannis and E. Stefanidaki, Min. and Met. J. **39**, 209 (2003)
5. M.F. Chambers, J.E. Murphy, BUREAU of MINES Report of Investigations, 9391 (1991)
6. S. Singh, J. M. Juneja, D. K. Bose, Appl. Electrochem. J. V **25**, 1139 (1995)
7. S. Jiao, H. Zhu, Hazard. Mater. J. V **189**, 821 (2011)

Properties of Polydisperse Tin-doped Dysprosium and Indium Oxides

Tatyana Malinovskaya[1], *Valentin* Suslyaev[2], *Valentina* Zhek[1], *Ilya* Lysak[1,3,*], *Sergey* Melentyev[4], *Kirill* Dorozhkin[2], *Eugeny* Korovin[2], and *Sergey* Pavlov[4]

[1] Siberian Physical Technical Institute at Tomsk State University, Tomsk, Russian Federation
[2] National Research Tomsk State University, Tomsk, Russian Federation
[3] National Research Tomsk Polytechnic University, Tomsk, Russian Federation
[4] Tomsk State University of Architecture and Building, Tomsk, Russian Federation

Abstract. The results of investigations of the complex permittivity, diffuse-reflectance, and characteristics of crystal lattices of tin-doped indium and dysprosium oxides are presented. Using the methods of spectroscopy and X-ray diffraction analysis, it is shown that doping of indium oxide with tin results in a significant increase of the components of the indium oxide complex permittivity and an appearance of the plasma resonance in its diffuse-reflectance spectra. This indicates the appearance of charge carriers with the concentration of more than 10^{21} cm^{-3} in the materials. On the other hand, doping of the dysprosium oxide with the same amount of tin has no effect on its optical and electromagnetic properties.

1 Introduction

Development of modern electronics leads to the need of searching for new promising materials, which are characterised by high chemical and thermal resistance, high permittivity (ε = 8-20) and resistivity (ρ = 10^{13}-10^{16} $\Omega\cdot$cm). Among them are rare-earth elements oxides (REO). Doping of oxides with various chemical elements allows modifying these parameters over a wide range. As an important material for electronics and optics, dysprosium oxide is also used for the manufacture of optical and magnetic memory and ultrapowerful magnets, as a neutron-absorbing material in nuclear power industry, as a component of red-emitting phosphors, and as a component of catalytic systems [1]. Therefore, studies of the dysprosium oxide doping with various elements are of considerable practical interest. The use of such dopants as Ce, Rh, Zr, and Hf [2-5] is well known, but there are practically no data on doping of dysprosium oxide with Sn. At the same time, the most widely studied oxide material used in electronics is indium oxide doped with tin (ITO), which is a solid solution of indium oxide (III) and tin (IV). An optimal weight ratio is 90% In_2O_3 and 10% SnO_2 by [6]. Doping with tin allows increasing concentration of charge carriers in indium oxide up to 10^{21} cm^{-3} due to the shallow impurity levels formed in its energy-band structure, which significantly affects the electrical and other properties of In_2O_3.

Thus, the aim of this work was to study the effect of tin-doping on the properties of indium and dysprosium oxides.

2 Experimental part

For the analysis of samples of oxides, the X-ray diffraction method was used. Investigations were carried out on a diffractometer Rigaku Miniflex 600 (USA) using CuKα-radiation in the range of 10°-90° (2θ) with the scanning step of 0.02° and rate of 2°/min. The identification of the diffraction peaks was performed using JCPDS database. The spectral dependences of the samples diffuse reflectance were recorded using a spectrophotometer IKS-31 (Russia) with photometric sphere and hemisphere. The spectra of complex permittivity (CP) in the frequency range of 115-720 GHz were measured using the "free space" method with a terahertz spectrometer STD-21 (Russia) [7]. Three backward-wave tubes were used to cover the entire range: 115-258, 220-511, and 430-720 GHz. Measurements were conducted on the 0.26 mm thick planar samples obtained by introducing the oxide powders in a bonding matrix. As the binder, a two-component polyurethane varnish brand Kontracid D3010 was used, the rationale for the choice of which and the coating manufacturing technique based on it are described in [8].

3 Results and discussion

It was found that the most complete coprecipitation of dysprosium and tin from theirs nitrate solutions takes place at pH values of 7-10, and that of indium and tin – at pH values of 6-10 (Figure 1). Thus, for doping of dysprosium and indium oxides with tin, we used the method of chemical coprecipitation of dysprosium or indium hydroxides coupled with tin hydroxide from the

* Corresponding author: lysak_ia@mail.com

nitrate solutions at pH7 with subsequent heat treatment in the air at 1000 °C for 1 hour. The tin content was 10 wt.%. As the precipitant, 25% ammonia water was used.

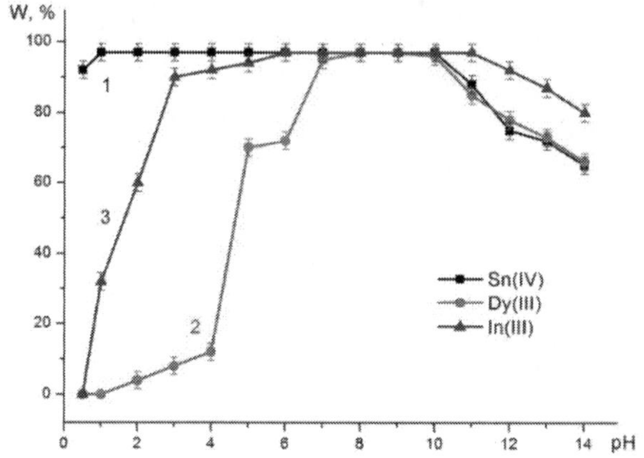

Fig. 1. Dependence of the deposition degree (*W*, %) with an aqueous solution of ammonia on the pH value of the nitrate solution: 1 – Sn(IV); 2 – Dy(III); 3 – In(III)

Diffractograms of ITO powders synthesized as described above were studied in [9]. The studies indicated that after the heat treatment of the coprecipitated indium and tin hydroxides for 1 hour in the air at 1000 °C, the formation of an In_2O_3 structure of cubic modification with the increased, compared to the standard values, interplanar distances was observed. Diffraction patterns of powders obtained as a result of heat treatment of dysprosium hydroxide and of coprecipitated dysprosium and tin hydroxides are presented in Figures 2 and 3, respectively. Comparison of the diffraction patterns with the standard data from the database JSPDS presented in these figures in the form of bar graphs suggests that the investigated powders are the dysprosium oxides of cubic modification with the lattice constants 1.0655767 nm and 1.0640638 nm for the undoped and Sn doped Dy_2O_3, respectively. Consequently, we can speak only about a slight decrease in the lattice parameter of dysprosium oxide doped with tin and assume the absence of the replacement of the dysprosium ions by tin ones. Studies of the complex permittivity spectra of the doped and undoped dysprosium oxide samples confirmed this assumption. As follows from the spectra analysis represented in Figures 4 and 5, the samples demonstrate a slight decrease of the real component and practically the constancy of the imaginary CP component in the entire frequency range. The scatter of values near 150 GHz, 220 GHz, and 450 GHz is most likely associated with an unstable work of each backward-wave tube at the beginning of its operating range.

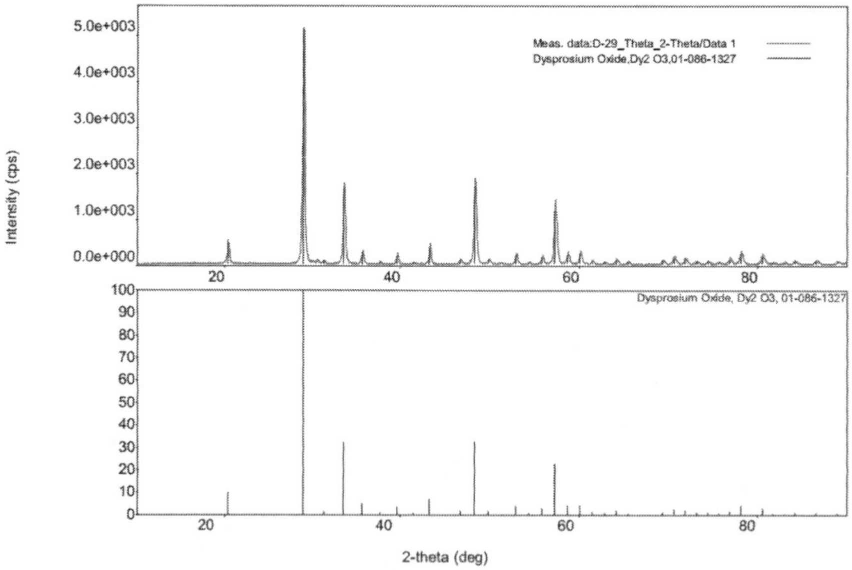

Fig. 2. X-Ray diffraction pattern of the powder obtained by heat treatment at 1000 °C for 1 hour of dysprosium hydroxide precipitated by aqueous ammonia from the nitrate solution at pH7.

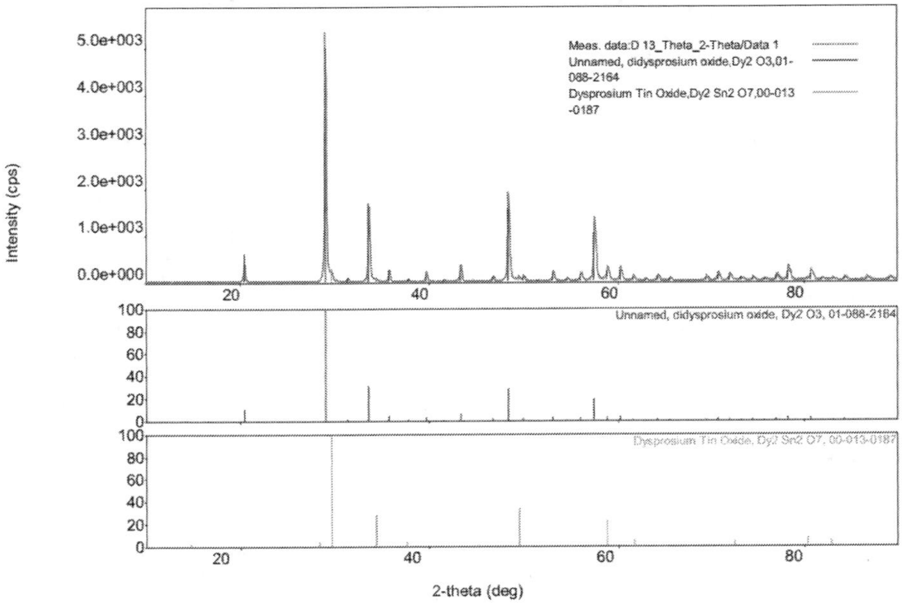

Fig. 3. X-Ray diffraction pattern of the powder obtained by heat treatment at 1000 °C for 1 hour of dysprosium and tin hydroxides coprecipitated by aqueous ammonia from the nitrate solutions at pH7.

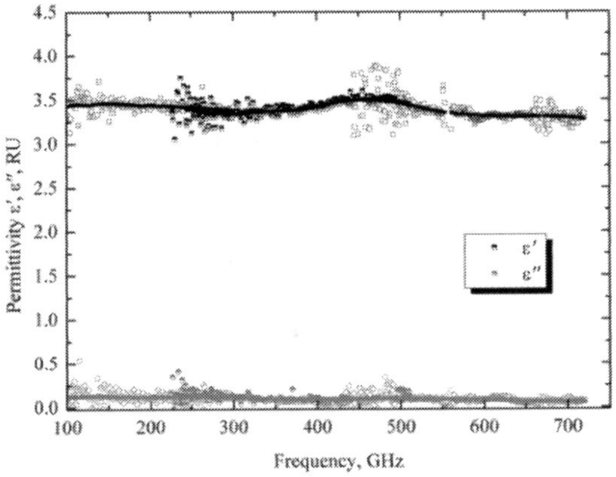

Fig. 4. Complex permittivity spectra of the dysprosium oxide samples.

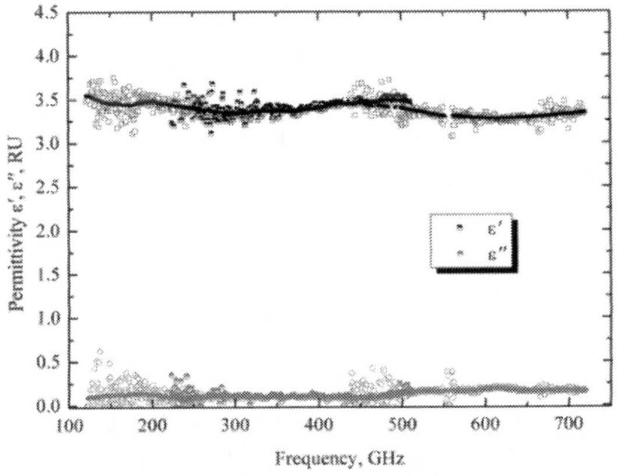

Fig. 5. Complex permittivity spectra of tin-doped dysprosium oxide samples.

This behavior of CP provides evidence that the studied materials exhibit electronic polarization, which is typical for dielectrics. No differences between the samples of pure dysprosium oxide and those doped with tin are detected, as opposed to indium oxide doped with tin, for which a sharp decrease in the real part of CP and high values of the imaginary part of CP are observed at low frequencies [10] indicating the presence of a conductive fraction - free electrons. High values of the CP components are also retained at high frequencies (Figure 6).

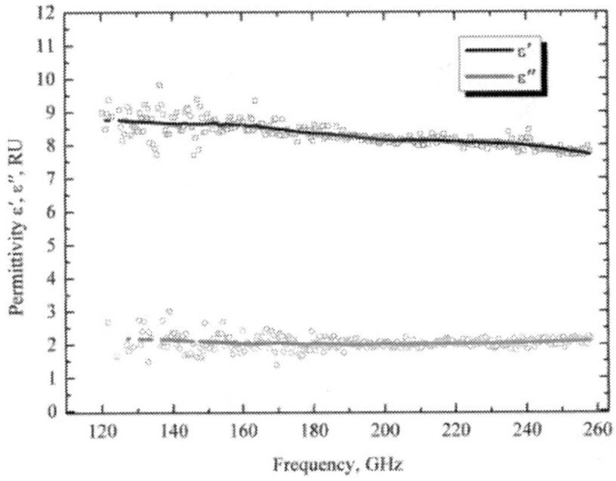

Fig. 6. Complex permittivity spectra of tin-doped indium oxide samples.

These differences affect the components of the electromagnetic response: transmittance T, reflectance R, and absorption factor A related by the formula $1 = R + T + A$. These components define the scope of practical applications of the materials under the study, such as screening devices to reduce the level of electromagnetic radiation that is necessary to solve the problems of electromagnetic compatibility, reduction of radio visibility, protection of biological objects, and creating anechoic chambers.

Experimental study of the electromagnetic response components (Figures 7 and 8) shows that from 65 to 80% of the incoming electromagnetic energy is propagated through the samples of pure and tin-doped dysprosium oxide with a very weak absorption, i.e. the screening properties are determined by the reflectance of samples. On the other hand, the doped indium oxide has a greater absorption of electromagnetic energy that is converted into the heat by the movement of free electrons caused by the electromagnetic field. The transmission factor decreases from 0.3 to 0.15 in the selected frequency range at relatively low reflectance values (Figure 9).

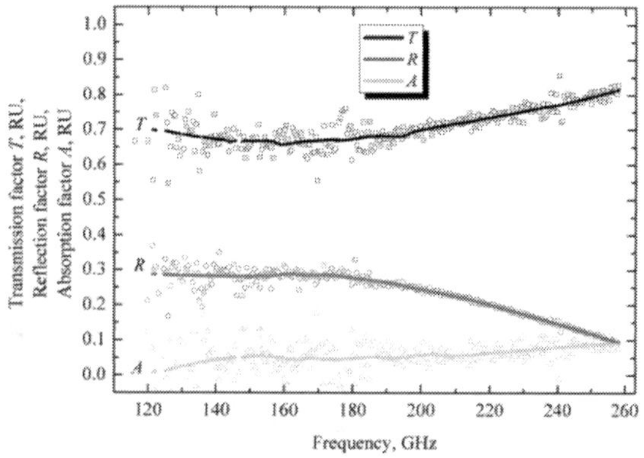

Fig. 7. Spectra of the electromagnetic response components of dysprosium oxide samples: *T* – transmission factor; *R* – reflection factor; *A* – absorption factor.

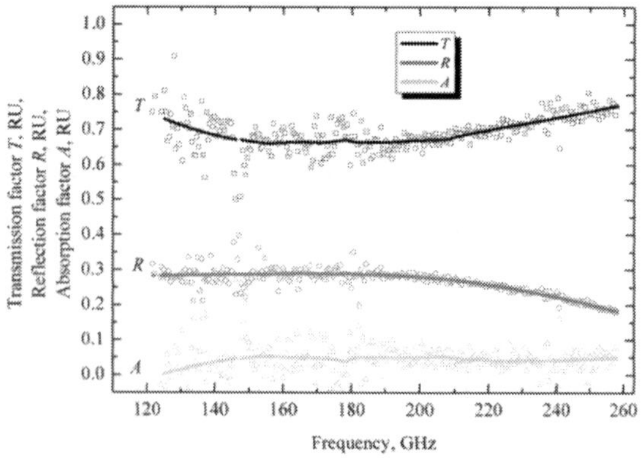

Fig. 8. Spectra of the electromagnetic response components of tin-doped dysprosium oxide samples: *T* – transmission factor; *R* – reflection factor; *A* – absorption factor.

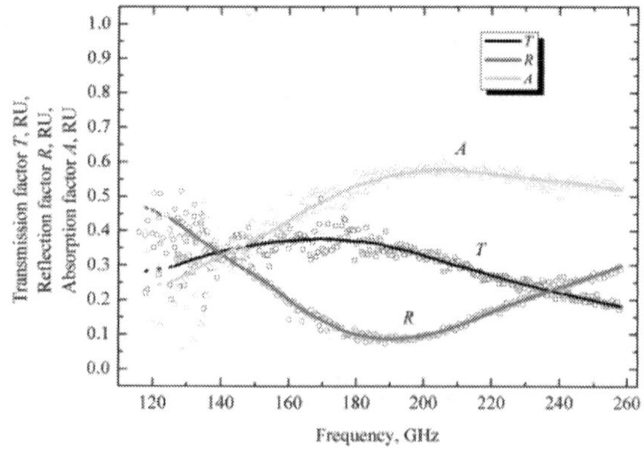

Fig. 9. Spectra of the electromagnetic response components of tin-doped indium oxide samples: *T* – transmission factor; *R* – reflection factor; *A* – absorption factor.

The absence of free electrons in dysprosium oxide doped with tin is confirmed by the absence of the plasma resonance in the spectral curves of the diffuse reflection factor of the tin-doped dysprosium oxide powder. This resonance is, however, observed in tin-doped indium oxide (Figure 10).

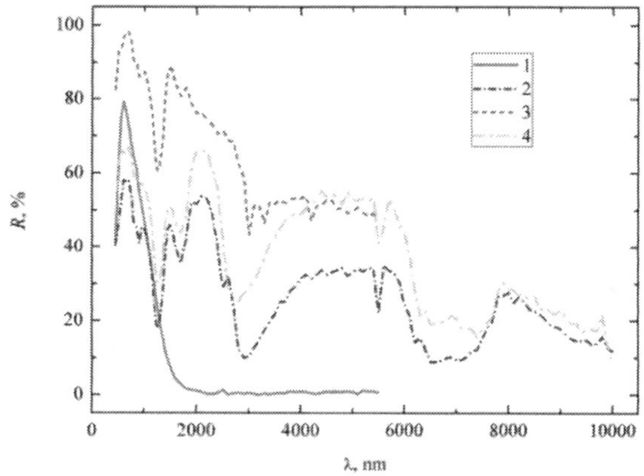

Fig. 10. Spectra of the diffuse reflection of powders of tin-doped indium oxide (1), tin-doped dysprosium oxide (2), pure indium oxide (3), and pure dysprosium oxide (4).

4 Summary and conclusion

Thus, doping of dysprosium oxide with tin has no effect on the electromagnetic and optical characteristics of the material. At the same time, doping of indium oxide with the same amount of tin results in a significant increase of the CP components and an appearance of the plasma resonance in its diffuse reflection spectra. The data obtained indicate the appearance of charge carriers with the concentration of 10^{21} cm^{-3} in the materials based on the dispersed ITO with the tin content of 10 wt. %.

The studies showed that the composite material based on the dispersed ITO doped with tin effectively interacts with the EHF electromagnetic radiation and can be used in the screening devices reducing the level of electromagnetic radiation, which is necessary for solving the problems of electromagnetic compatibility, reduction of radio visibility, protection of biological objects, and creating the anechoic chambers.

Acknowledgments

The work was supported by the Tomsk State University Competitiveness Improvement Programme, registration No. 8.1.23.2015.

References

1. H. S. Nalwa, *Handbook of Nanostructured Materials and Nanotechnology* (Academia Press, San Diego, 1999).
2. M. Mikuśkiewicz, M. Stopyra, G. Moskal, Arch. Metall. Mater., **61, 2B**, 965–969 (2016).
3. R. C. Layland, S. L. Kirkland, H. C. Loye, J. Solid State Chem., **139, 1** p. 79–84 (1998).
4. V. D. Risovany, A. V. Zakharov, E. M. Muraleva, et al., J. Nucl. Mater. **355, 1**, 163–170 (2006).
5. Q. Xu, W. Pan, J. Wang, et al., J. Amer. Ceram. Soc. **89, 1**, 340–342 (2006).
6. D. S. Ginley, *Handbook of transparent conductors* (Springer, New York, 2010).
7. V. I. Suslyaev, G. E. Dunaevskyi, E. V. Emelyanov, G. E. Kuleshov, Proceedings of the higher educational institutions. Physics (in Russian – Izvestiya vuzov. Physica), **54, 9/2**, 138–146 (2011).
8. T. D. Malinovskaya, V. I. Suslyaev, S. V. Melentyev, K. V. Dorozkin, Russ. Phys. J. **57, 8**, 1094–1098 (2014).
9. Y. P. Egorov, T. D. Malinovskaya, E. P. Naiden, V. I. Sachkov, E. I. Sachkova, Chem. Sustain. Dev. **10**, 679-685 (2002).
10. T. D. Malinovskaya, V. I. Suslyaev, V. A. Zhuravlev, S. V. Melentyev, E. Yu. Korovin, K. V. Dorozhkin, V. V. Zhek, S. V. Pavlov, Proceedings of the higher educational institutions. Physics (in Russian – Izvestiya vuzov. Physica), **59, 9**, 168–169 (2016).

Cement-free Binders for Radioactive Waste Produced from Blast-furnace Slag using Vortex Layer Activation Technology

Ilya Mazov[1,2], *Bekzod* Khaydarov[1,*], *Dmitrii* Suvorov[1], *Tamara* Yudintseva[1], *Denis Kuznetsov*[1], *Andrey* Yudin[1]

[1] National University of Science and Technology "MISIS", Moscow, Russian Federation
[2] National Research Tomsk State University, Tomsk, Russian Federation

Abstract. The paper addresses the issue of recycling granulated blast-furnace slag (gBFS) as a source for production of cement-free binder materials for further usage in rare-earth metals production for radioactive waste disposal. The use of the vortex layer activator was provided as main technique allowing to produce high-dispersed chemically activated binders. The paper examines the effect of processing conditions on the physical-chemical and mechanical properties of the resulting BFS-based cement-free materials and gBFS-based concretes.

1 Introduction

Radioactive waste is a major problem in the production of rare-earth metals. Burial thorium and its products are extremely important. The easiest way - is cementing slurry containing all the radionuclides, which are contained in the feedstock, with burial in earthen trenches. The disadvantage of this method is significantly higher consumption of expensive cement. The use of waste iron and steel industry may reduce the cost of the process [1].

Typical industrial metallurgical production results in formation of large amount of technological wastes – ash, slag etc. Steelmaking produces as much slag as 200-1000 kg per ton of steel. Large amounts of steel slag are wasting soil and water, resulting in high threat to natural resources. Typical yearly production of steel slag in Russia is 5-7 Mton, which makes this problem very important [2].

Blast-furnace slag (BFS) is usually treated in molten state, producing so-called granulated blast-furnace slag (gBFS), containing 3-5 wt. % of metal inclusions [3]. Granulated BFS usually contains certain amount of alumina, silica and calcium oxide, thus making it perspective source for production of construction materials. Production of high quality cement-free binders using gBFS as a source shows several advantages – such as reduction of ecological threats, usage of local source for road and civilian building etc [4, 5].

One of the most difficult problems in this area is improvement of energy effectiveness of the production process of mechanically activated high-dispersive materials, which are needed for production of high-strength binders [6, 7]. Traditional methods, such as hummer mills or ball milling are characterized by high energy consumption and periodical process [8].

The main aim of the work is developing of energy efficient technology for production of clinker-free materials (CFM) and CFM-based concretes by vortex layer mechanical activation using gBFS as source materials [9] and investigation of mechanical properties of produced materials depending on preparation conditions.

2 Methods and materials

Granulated BFS was produced by JSC "Severstal" (Russia). This type of BFS is typical for most large (producing more than 3 Mtons of metal per year) metallurgical companies of Russia. Commercial sample of the Portland cement M400 was used as comparative sample.

All chemicals used in this work were purchased locally and are reagent grade or higher.

Investigation of phase and chemical composition was performed using X-ray fluorescent spectrometer ARL 9900 (Thermo Fisher Scientific, USA), providing standard-free analysis.

Particle size distribution was investigated using Analysette 22 NanoTec (Fritsch, Germany). Mean particle size and size distribution was calculated using standard integrated software of the analyzer.

Specific surface area of all samples was measured using Quantachrome NOVA 1200 Surface Area Analyzer (Quantachrome Instruments, USA).

Phase composition of gBFS (both pristine and chemically modified) and corresponding concretes (after natural hardening and hydro thermal treatment (HTT)) were investigated using XRD method. X-ray diffraction measurements were performed using X-ray diffractometer Difray 401 (Scientific Instruments, Russia).

Mechanical properties of concretes were investigated for cubic samples with dimensions 2×2×2 cm^3 using Instron 150 LX universal testing machine (Instron, USA).

* Corresponding author: bekzod1991@mail.ru

Vortex-layer activator (VLA) was used for mechanical and chemical activation of slag. VLA setup consists of electromagnetic coil (inductor) placed in water-cooling hull. Stainless steel beaker is placed inside the coil and filled with material to treat and magnetic (steel) milling elements. In this work we used stainless steel cylinder with size Ø2×15 mm^3. VLA setup allows to perform continuous treatment and is easy scalable.

Two main parameters of the activation process determine properties of the slag, activated using VLA apparatus. These are treatment time and "slag:grinder" ratio. Typical physical-chemical properties such as specific surface area, particle size distribution and morphology of particles were investigated under variation of treatment parameters. Comparative study of properties was performed for pristine commercial Portland cement sample.

Samples of gBFS-based concretes were prepared using mechanically activated slag. Chemically modified samples were prepared using NaOH. Solid base (7 wt.%) was added to slag before mechanical activation.

Comparative study was performed using concrete samples prepared by mixing of slag or Portland cement with water, typical "water/binder" ratio was set to 0.22. Note that all concrete samples were investigated after 28 days of hardening.

3 Results and Discussion

3.1 Chemical and Phase Composition of gBFS

Chemical composition of pristine gBFS is shown in the Table 1. According to chemical composition of the slag its main parameters were calculated, allowing to conclude that the slag sample is mainly basic (Base module MB = 1.19, Base coefficient KB = 1.15, Activity Index MA = 3.19, Quality coefficient KQ = 1.59).

Table 1. Chemical composition of gBFS.

Compound	Amount, at. %
Na_2O	0.28
MgO	12.18
Al_2O_3	8.91
SiO_2	34.85
P_2O_5	0.01
K_2O	0.43
CaO	40.03
TiO_2	1.06
MnO	0.12
Fe_2O_3	1.22

As one can see from the Table 1, the content of inactive impurities (iron and manganese oxides) is low.

Chemical and phase composition of gBFS corresponds to that of typical cement clinker materials, allowing one to use it as a source for preparation of clinker-free binders.

3.2 Influence of activation conditions on slag properties

Results of particle size and SSA (specific surface area) measurement for gBFS, treated in various conditions, are shown in Table 2. Typical SEM images of activated samples are shown on the Fig. 1.

Table 2. Mean size and specific surface area of gBFS treated in various conditions.

"slag/grinder" ratio	Treatment time					
	3 min		5 min		7 min	
	Particle size, µm	SSA, m^2/g	Particle size, µm	SSA, m^2/g	Particle size, µm	SSA, m^2/g
1:2	10.8	1.509	8.0	2.244	8.5	3.267
1:2.5	27.6	1.454	16.7	2.256	9.7	2.154
1:3	38.9	1.39	26.6	1.362	14.2	2.489
1:3.5	47.3	0.61	10.9	1.379	22.6	1.917

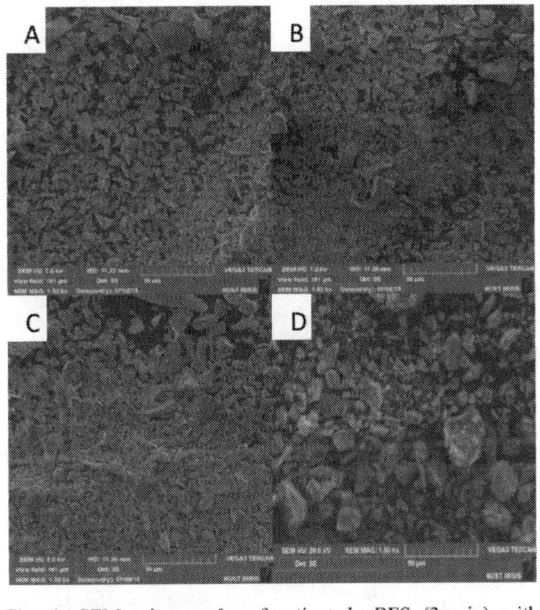

Fig. 1. SEM micrographs of activated gBFS (3 min) with different "slag/grinder" ratio (A – 1:2, B – 1:3, C – 1:3.5, D – pristine Portland cement).

According to the data of Table 2 and Fig. 1 it is possible to conclude that increase of the amount of grinding media in VLA setup results in increase of the mean particle size and in lowering of the specific surface area. This phenomenon most probably occurs due to space limitations in activation chamber and following decrease of the activation efficiency and agglomeration of the slag particles. This is due to increase of number of grinder-to-grinder impacts instead of grinder-to-slag impacts. Thus optimal "slag/grinder" ratio is 1:2…1:2.5.

Increase of the treatment period results in simultaneous increase of the specific surface area and diminishing of the slag particles.

Taking into account above mentioned considerations, following conditions of VLA treatment were chosen as optimal: 5 minutes treatment time with "slag/grinder" ratio 1:2. Treatment in such conditions allows to produce activated gBFS with particle size 10-15 μm and SSA = 2.2…3.2 m^2/g.

Properties of the commercial sample of Portland cement were also investigated. Typical SSA value is 1.7…1.9 m^2/g and mean particle size is 14…16 μm. Thus mechanically activated gBFS shows structural properties close to that of Portland cement and the main issue, determining its possible application as a clinker-free binder, is phase composition of pristine materials and corresponding concretes.

3.3 Phase composition of gBFS and corresponding concretes

XRD data for both Portland cement and PC-based concrete shows high amount of alite (C3S) and belite (C2S) phases (Fig.2). In contrast, pristine g-BFS is almost amorphous and no significant peaks can be found on XRD spectrum.

Certain amount of crystal (or semi-crystal) alite phase can be found in VLA-treated gBFS sample. The same peak (of hydrated alite phase) can be found on the X-ray diffractogram of gBFS-based concrete. However, amount of this phase is relatively low and it should be raised in order to improve mechanical properties of concrete.

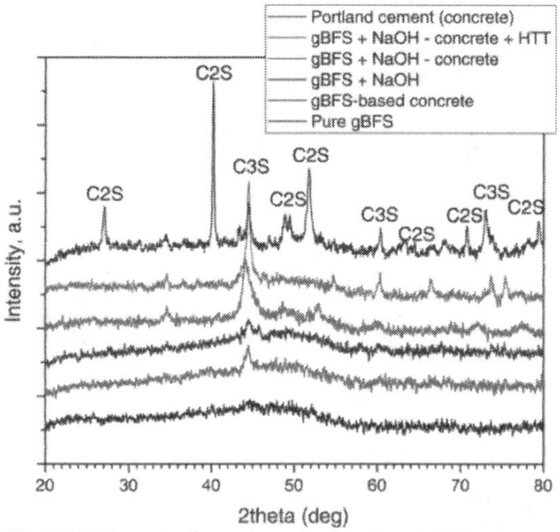

Fig. 2. XRD spectra for various slag-based materials.

Addition of NaOH to the gBFS during VLA treatment results in significant increase of amount of crystal phases both in slag and slag-based concrete. Still amount of the crystal phase is low as compared with Portland cement and PC-based concrete.

Hydrothermal treatment of NaOH-modified gBFS-based concrete results in further increase of the amount of alite and belite phases thereby it is comparable with that of PC-based concrete.

3.4 Mechanical properties of concretes

Results of investigation of mechanical properties of PC-based and gBFS-based concretes are shown in the Table 3. According to the presented data it can conclude that mechanical properties of gBFS-based concretes are strictly dependent on phase composition and on mechanical and chemical activation as well as on post-hardening treatment of the concrete.

The highest strength value was achieved for the sample of NaOH-activated g-BFS which was hydrothermally treated after hardening. This sample shows the highest amount of crystal phases of XRD spectra and is comparable with PC sample.

Table 3. Mechanical properties of PC-based and gBFS-based concretes.

Sample	Strength
PC M400	39.5 MPa
gBFS w/o chemical modification	12.1 MPa
gBFS + 7% NaOH w/o post-treatment	38.7 MPa
gBFS + 7% NaOH + HT treatment	44.7 MPa

Energy efficiency of VLA approach was calculated in comparison with traditional ball milling technology. In order to produce activated gBFS with high specific surface area (1.5-3.2 m^2/g) one needs to apply VLA activation for 3-7 minutes, and at the same time ball milling approach needs several hours to achieve same parameters. Energy consumption depends strongly on the power of the mill used and can be estimated as 3-5 times higher for ball mill as compared with VLA setup.

4 Conclusion

Clinker-free cementitious binders were prepared using granulated blastfurnace slag as a source material. Novel vortex layer activation approach was used for preparation of mechanically and chemically activated materials.

Structural and mechanical properties of slag-based materials were investigated. Such properties' depending on preparation conditions as well as optimal activation conditions were determined.

Mechanical properties of slag-based concretes were investigated depending on preparation conditions and post-hardening treatment. It was shown that strength of NaOH-activated slag-based concretes with hydrothermal treatment is outreaching that of PC-based concretes (44.7 MPa and 39.5 MPa, correspondingly). It makes slag-based materials perspective for burial of rare-earth and radioactive wastes.

Presented work indicate that the developed method of clinker-free cement producing may be suitable for the production of binder materials for the recycling and burial of rare and radioactive earth metals waste disposal. The use of this material will simplify the technology and reduce costs due to the use of metallurgical production waste.

Authors are kindly grateful for support of this work by Ministry of Science and Education of Russian Federation (project RFMEFI57814X0099).

References

1. R. Ausness, Wis. L. Rev. **707** (1979)
2. J. Zelic, Cement Concrete Res. **35,** 2340 (2005)
3. S. Bostancı, M. Limbachiya, H. Kew. J. Cleaner Production **112**(1), 542 (2016)
4. W.T. Kuo, H.Y. Wang, C.Y. Shu. Construction and Build. Mater., **63**, 189 (2014)
5. I. Ismail, A. S. Bernal, J. L. Provis, e.a. Cement and Concrete Composites, **45**, p. 125–135 (2014)
6. A.S. Bernal, R.M. de Gutiérrez, L.J. Provis, Construction and Build. Mater., **33**, 99 (2012)
7. X. Huang, Z. Wang, Y. Liu, W. Hu, W. Ni. Construction and Build. Mater. **112**, 241 (2016)
8. J.I.E. Garcia, J.M. Nonell, A. Gorokhovsky, Bol. Soc. Esp. Ceram. Vidrio, **41**(5), 451 (2002)
9. A.V. Gorokhovsky, J.I. Escalante-Garcia, E. Sanchez-Valdes, I.N. Burmistrov, D.V. Kuznetsov, Ceramics Int. **41**, 13294 (2015)

Excitation Frequencies of the Effects of Selective Drift of Solvated Cations in Moving Salts Solution

Igor Shamanin[1], *Mishik* Kazaryan[2,3], *Valeriy* Gofman[1], *Sergey* Timchenko[1,*], *Andrei* Poberezhnikov[1], *Ilya* Tuksov[1]

[1] National Research Tomsk Polytechnic University, Tomsk, Russian Federation
[2] Lebedev Physical Institute of the Russian Academy of Sciences, Moscow, Russian Federation
[3] National Research Tomsk state University, Tomsk, Russian Federation

Abstract. Interaction of external electric field with free and bound charges in the salt solution volume in polar dielectric fluid was considered and the equation describing ion oscillations relative to solvent molecules was obtained. Fundamental frequencies of the solvation shell relative to the ion in different approximations describing the system: ion - solvation shell were evaluated. The results of comparison with the experiment pointed out that the model, in which the solvated ion is considered as a spherical rotator formed by the solvent molecules layer bordering with the external solvation shell, is the most appropriate. It provides an opportunity of development of new technologies for extraction of rare earth elements.

1 Introduction

Experimental data and theoretical estimations [1] have shown, that excitation of rotary-onward motions of solvated cations (clusters) with various inertial properties can become a basis for technology of separation of rare earth elements in water solutions of their salts. Practical use of the discovered phenomenon of metals solvated cations selective field-assisted motion in solutions of salt mixtures, influenced by external periodic electric field, will require the presence of some relations applicable for engineering calculations. In particular, to determine velocity of motion of the solvated ion under direct influence of external electrical field the parameters of its oscillatory motion caused by that field need to be defined. The solvated ion is in a polar dielectric fluid and the electric field influences both the ion and the solvent molecules. In its turn, the internal electric field influences the solvent molecules surrounding the ion. The task of determining the equation describing ion oscillations relating to the surrounding solvent molecules is not trivial.

To determine such equation, interrelation of electric fields and currents corresponding to some charge distribution in the salt solution in polar dielectric fluid should be considering.

2 External electric field influences on free and bound charges in solution

Electric field intensity distribution $\vec{E}(\vec{r})$ and volume charge density distribution $\rho(\vec{r})$ are connected by Poisson's equation [2]:

$$div\vec{E}(\vec{r}) = 4\pi\rho(\vec{r}).$$

If salt, the molecules of which are dissociated, is dissolved in dielectric fluid, the electric field in the fluid volume is created by free charge carriers – ions formed at dissociation with the volume charge density distribution $\rho_f(\vec{r})$. This field intensity distribution is also described by Poisson's equation:

$$div\vec{E}_f(\vec{r}) = 4\pi\rho_f(\vec{r}).$$

The molecules of dielectric fluid (solvent) polarize in the electric field of free charge carriers. So the following condition is fulfilled [3]:

$$div(\vec{E}(\vec{r}) + 4\pi\vec{P}(\vec{r})) = 4\pi\rho(\vec{r}), \quad (1)$$

where $\vec{P}(\vec{r})$ - polarization density distribution. Moreover, this condition is fulfilled, no matter how $\vec{E}(\vec{r})$ and $\vec{P}(\vec{r})$ is connected with each other.

When the external electric field is applied to the solution, it results in its local electrical neutrality breaking, that is, volume charge density other than zero appears. Besides, these charges are connected and appear entirely due to non-uniform polarization of dielectric fluid molecules. Volume density distribution of the bound charges $\rho_b(\vec{r})$ is determined by the relation:

$$\rho_b(\vec{r}) = -div\vec{P}(\vec{r}). \quad (2)$$

In the general case the volume charge consists of two components – free charges density and bound charge density, which appear when there is non-uniform polarization of the sample $\vec{P}(\vec{r})$.

Consideration of one-dimensional case (when the solution is in the form of a flat slab, the transverse sizes of which are much greater than its thickness) can be confined for simplicity. The polarization density is

* Corresponding author: timsn@tpu.ru

connected with the external electric field intensity $\vec{E}_0$ created by plan-parallel electrodes isolated from the solution by the relation $4\pi\vec{P} = \vec{E}_0 - \vec{E}$,

where $\vec{E}$ is the field intensity inside the layer of the solution located between the electrodes.

The electric field intensity inside the solution is distributed in a particular way, that is, it depends on the coordinate. At that, the averaged intensity value in the solution volume $\langle E\rangle$ is smaller than the filed intensity between the electrodes E_0. The ratio $E_0/\langle E\rangle$ can be evaluated using the equation (1) and the correlation between the polarization density and intensity of the field causing dielectric fluid polarization:

$$\vec{P} = \frac{3}{4\pi}\left(\frac{\varepsilon-1}{\varepsilon+2}\right)\vec{E},$$

where ε - dielectric permeability of the solvent.

If the electric field is created by the charge with the volume density ρ, and polarization is caused by the electric field of this charge, the equation (1) can be rewritten in the following form:

$$\left(1+3\left(\frac{\varepsilon-1}{\varepsilon+2}\right)\right) div\vec{E} = 4\pi\rho.$$

This equation shows that when the charge density ρ is the same, polarization causes $\left(1+3\left(\frac{\varepsilon-1}{\varepsilon+2}\right)\right)$ times decrease of the electric field intensity. When dielectric permeability has greater values $(\varepsilon > 10)$, the field intensity in the dielectric fluid volume $\langle E\rangle$ is approximately 4 times smaller than the intensity between the electrodes E_0. When the values of dielectric permeability are close to 1, the field intensity in the dielectric fluid volume $\langle E\rangle$ is approximately 2 times smaller than the intensity between the electrodes E_0.

If the electric field intensity distribution is considered, it follows from the mentioned above that the intensity increases in the plane solution layer volume from the value $\langle E\rangle$ to the value E_0 at the approach to the electrodes isolated from the solution. Macroscopic value of the bound charges density $\rho_b(\vec{r})$ will be not equal to zero in these solution regions close to the electrodes, because moving from the electrodes the field intensity will be weakly depend on the coordinate and will have the value $\langle E\rangle \approx const$. Also the polarization density $\vec{P}$ will not change significantly and $div\vec{P}(\vec{r})$ will be close to zero. Consequently, the macroscopic value of bound charges density $\rho_b(\vec{r})$ by virtue of the equation (2), tends to zero while moving away from the electrodes.

Taking into consideration the above presented reasoning it is possible to write the equation connecting electric fields intensities in the solution volume with the free charge density in the saline solution in the dielectric fluid:

$$div\left(\vec{E}_f(\vec{r}) + 3\left(\frac{\varepsilon-1}{\varepsilon+2}\right)\left(\vec{E}_0(\vec{r}) + \vec{E}_f(\vec{r})\right)\right) = 4\pi\rho_f(\vec{r}).$$

The equation for the solution layer which is between two plane electrodes isolated from the solution will have the following form:

$$\frac{\partial}{\partial x}\left(E_f(x) + 3\left(\frac{\varepsilon-1}{\varepsilon+2}\right)\left(E_0(x) + E_f(x)\right)\right) = 4\pi\rho_f(x),$$

or, which is the same, as:

$$\left(1+3\left(\frac{\varepsilon-1}{\varepsilon+2}\right)\right)\frac{\partial E_f(x)}{\partial x} + 3\left(\frac{\varepsilon-1}{\varepsilon+2}\right)\frac{\partial E_0(x)}{\partial x} = 4\pi\rho_f(x).$$

3 Ion oscillation equation relative to solvent molecules

Differentiate the obtained equation with respect to time:

$$\left(1+3\left(\frac{\varepsilon-1}{\varepsilon+2}\right)\right)\frac{\partial}{\partial t}\left(\frac{\partial E_f(x)}{\partial x}\right) + 3\left(\frac{\varepsilon-1}{\varepsilon+2}\right)\frac{\partial}{\partial t}\left(\frac{\partial E_0(x)}{\partial x}\right) = 4\pi\frac{\partial\rho_f(x)}{\partial t}. \quad (3)$$

The right member of this equation is none other than the current density j_{ion} of free charge carriers (ions) multiplied by 4π:

$$4\pi\frac{\partial\rho_f(x)}{\partial t} = 4\pi j_{ion}(t).$$

For a single ion located in the solution the space intensity distribution of the electric field created by it can be defined in a functional form:

$$E_f(x) = kq_{ion}\frac{x - x_0(t)}{|x - x_0(t)|^3},$$

where k - invariable, q_{ion} - ion electric charge, $x_0(t)$ - ion coordinate, which changes with respect to time due to ion motion in the external electric field. Then:

$$\frac{\partial}{\partial t}\left(\frac{\partial E_f(x)}{\partial x}\right) = kq_{ion}\frac{\partial}{\partial t}\left(\frac{1}{|x - x_0(t)|^3} - 3\frac{1}{(x - x_0(t))^3}\right)$$

and

$$\frac{\partial}{\partial t}\left(\frac{\partial E_f(x)}{\partial x}\right) = -6\frac{kq_{ion}}{(x - x_0(t))^4}\frac{\partial x_0(t)}{\partial t} = -6\frac{kq_{ion}}{(x - x_0(t))^4}v_{ion}(t),$$

where $v_{ion}(t) = \frac{\partial x_0(t)}{\partial t}$ - the directed ions motion velocity.

Taking into consideration that the ions current density is $j_{ion}(t) = nq_{ion}v_{ion}(t)$, where n - the ions density, the equation (3) for the coordinate of the single ion $x_i(t)$, located in the salt solution in the polar dielectric fluid with great dielectric permeability, can be written as follows:

$$-24\frac{kq_{ion}}{(x - x_i(t))^4}\frac{\partial x_i(t)}{\partial t} + 3\frac{\partial}{\partial t}\left(\frac{\partial E_0(x)}{\partial x}\right) = 4\pi nq_{ion}\frac{\partial x_i(t)}{\partial t},$$

or

 DOI: 10.1051/matecconf/20179600012

$$\left(4\pi n + 24\frac{k}{(x-x_i(t))^4}\right)q_{ion}\frac{\partial x_i(t)}{\partial t} = 3\frac{\partial}{\partial t}\left(\frac{\partial E_0(x)}{\partial x}\right).$$

The numerical value of the invariable k (the proportionality factor in the coulomb law) in the system SI of physical quantities is $k \approx 9\cdot10^9$. Let us choose the fixed point in the solution volume $x = a$, in which the gradient of the field intensity is known, and its time dependence is $\left(\frac{\partial E_0(a)}{\partial x}\right) = f(t)$. In the experiments on excitation of the field-assisted motion of solvated ions [4] $f(t) \approx A(1+2\sin\omega t - \sin 2\omega t)$, the invariable value $A \approx 4\cdot10^6$ V·m^{-2}. The ion oscillation amplitude influenced by the external electric field $(a - x_i(t))_{max}$ certainly does not exceed 10^{-3} m, thus, the first part within the brackets of the left side of the equation can be neglected. Therefore, the equation of ion oscillation $\Delta(t)$ can be written as follows:

$$24kq_{ion}\frac{1}{\Delta(t)^4}\frac{\partial}{\partial t}\Delta(t) = 6A\omega\left(\cos\omega t - \cos 2\omega t\right),$$

or

$$\frac{1}{\Delta(t)^4}\frac{\partial}{\partial t}\Delta(t) = \frac{A\omega}{4kq_{ion}}\left(\cos\omega t - \cos 2\omega t\right). \qquad (4)$$

The solution of the equation (4) takes the form [5]:

$$\Delta(t) = \left(\frac{A}{8kq_{ion}}\left(\sin 2\omega t - 2\sin\omega t\right) + C\right)^{-1/3},$$

where the invariable C is determined from the condition $\Delta(t=0) = a$.

The oscillation frequency $\nu = \frac{\omega}{2\pi}$ is the same as the frequency of the external electric field and is the oscillation frequency of the single ion relative to the solvation shell. If this frequency is coincident with the fundamental frequency of the solvated ion (the system: the ion - the solvation shell), the transition of oscillative motion into translational field-assisted motion of the solvated ion should be expected.

4 Solvated ion fundamental frequencies

4.1 Solvated ion (cluster) as a liquid droplet

Consider the cluster formed by the central ion and the surrounding associated solvent molecules as a liquid droplet. The fundamental frequencies for the droplet oscillation relative to static equilibrium position on the assumption of the minuteness of velocities field perturbation inside the droplet and free surface can be defined by the formula [6]:

$$\nu_l^2 = \frac{\sigma_s}{\rho R_0^3}l(l-1)(l+2), l = 2,3,...,$$

where ρ - fluid density, σ_s - surface tension coefficient, R_0 - droplet (cluster) radius.

In order to define the surface tension coefficient it is necessary to evaluate the work to be done to create the surface limiting the cluster. Coming out of the solvent molecule to this surface leads to increase of the cluster energy by the binding energy of the solvent molecule inside the cluster. The surface tension coefficient will be equal to the ratio of this energy ε_b to the area of the sphere surface s, falling on one solvent molecule:

$$\sigma_s = \frac{\varepsilon_b}{s}.$$

The cluster radius is determined from the condition of equality of critical electric field intensity to the intensity value near the central ion [1]. In Figure 1 the distributions $E_r(r)$ for different ions in water are presented.

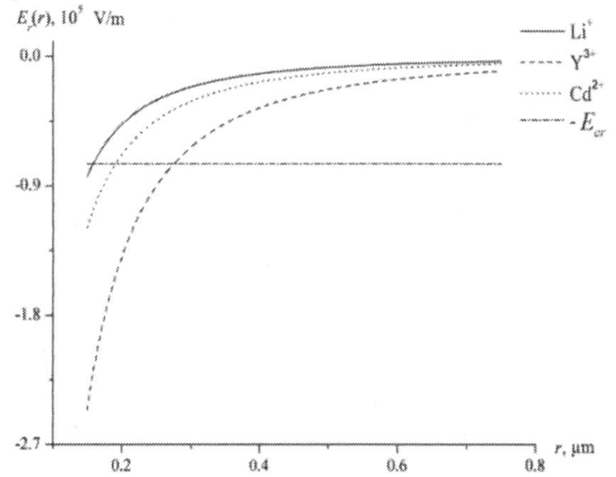

Fig. 1. Dependencies $E_r(r)$ for cations Li^+, Cd^{2+}, Y^{3+} for the temperature of 298 K and critical electric field intensity E_{cr}.

Consider that the cluster radius R_0 is equal to 0.2 μm ($E_r(r) = E_{cr}$) for evaluation. At usual conditions $9\cdot10^5$ molecules of water can be located within the sphere with the radius R_0. The sphere surface area is $50.24\cdot10^{-14}$ m^2, and $s \approx 5.55\cdot10^{-19}$ m^2 falls on one molecule.

To determine the binding energy of the solvent molecule inside the cluster it is possible to assume that it is equal to the dipole binding energy in the central ion electric field:

$$\varepsilon_b \sim p\cdot E,$$

where $p-$ intrinsic dipole moment of the solvent molecule, $E-$ electric field intensity within the cluster radius (see Figure 1). The binding energy ε_b for the intrinsic dipole moment of the water molecule p = $6.2\cdot10^{-30}$ C·m will be about 10^{-24} J. The surface tension coefficient will be $\sigma_s \approx 2\cdot10^{-6}$ J/m^2. Therefore, the intrinsic oscillation frequency of the cluster as a droplet can reach several hundreds of kHz. It is equal to the hydrogen binding energy in the solvent which falls on one solvent molecule. The hydrogen binding energy in water is 21 kJ/mole, consequently, $\varepsilon_b \approx 3.5\cdot10^{-20}$ J falls on one solvent molecule. The surface tension coefficient is $\sigma_s \approx 0.063$ J/m^2. Then the intrinsic oscillation frequency of the cluster as a droplet is tens of MHz.

4.2 Solvated ion (cluster) as a spherical pendulum

Consider the cluster formed by the central ion and the surrounding associated solvent molecules as a hollow sphere with the inner radius R_1, the outer radius R_2 and the mass M. In this case the hollow sphere can rotate about the axis passing through its geometric center. The sphere is connected with the axis by elastic spiral spring with the tension k, which provides the sphere with a steady state position. In this model the cluster is considered as a torsional spherical pendulum. The spring tension is determined by the binding energy of the central ion with the solvent molecules located at some distance from it. These molecules form the first solvation radius R_1, and their amount is equal to the ion coordination number in the solvent.

The fundamental oscillations frequency of such a pendulum is defined by the formula:

$$\nu = \sqrt{\frac{k}{J}},$$

where J - moment of pendulum inertia the value of which can be defined by the formulae:

$$J = \frac{2}{5}MR_2^2 \text{ - for } R_2 >> R_1 ;$$

$$J = \frac{2}{3}M\left(\frac{R_1+R_2}{2}\right)^2 \text{ - for } R_2 \sim R_1 .$$

The spring tension is defined from breakeven condition of electrostatic interaction of the central ion with polarized solvent molecules located at the first solvation radius, and the spring tension energy:

$$k(R_1 - r_i) = E \cdot n \cdot q ,$$

where r_i - central ion radius, n - solvent molecules amount, q - partial charge which is obtained by the solvent molecule in the ion electric field.

For the ion Ca^{2+} the radius r_i=0.114·10^{-9} m, when the coordination number is n=6, and r_i=0.148·10^{-9} m, when the coordination number is n=12. As this takes place, the first solvation radius will be R_1=0.229·10^{-8} m for n=6 and R_1=0.324·10^{-8} m for n=12. Supposing that the outer radius R_2 is determined from the condition $E_r(R_2) = E_{cr}$ and is equal to 0.2 μm, then, the mass of the spherical torsional pendulum will be $M \approx$33.5·10^{-18} kg. If the solvent is water, water molecules will be oriented in such a way that oxygen atoms will be located closer to the central ion Ca^{2+}. By virtue of significant difference in electronegativity of hydrogen and oxygen the oxygen atom in the water molecule acquires significant negative partial charge q= -0.66 e. As a result the tension value for the solvated ion Ca^{2+} will be k= (45…70)·10^{-6} N/m. So the natural oscillation frequency of the cluster as a torsional spherical pendulum will be about 10 THz.

4.3 Solvated ion (cluster) as a spherical rotator

Consider the cluster formed by the central ion and the surrounding associated solvent molecules as a hollow sphere with the inner radius R_1, outer radius R_2 ($R_2 >> R_1$) and the mass M. Insider the hollow spherical shell there is the central ion.

The sphere volume is rather small compared to the cluster volume having the radius R_2. However, the hollow sphere can rotate about the mass center of the cluster which practically aligns with the spherical shell geometrical center, because the shell mass M is significantly more than the central ion mass m_i. Therefore, natural frequencies for such a system can be defined by the formula:

$$\nu_l = \frac{h}{8\pi^2 J} l(l+1), l = 1,2,3,...,$$

where h - Planck's constant, $J = MR_2^2$ - inertia moment of the rotator.

Depending on the value R_2, which in its turn defines the value M, natural frequencies change from units of Hz for the cluster with the radius $R_2 \sim 0.1$ μm up to tens of Hz for the cluster the solvation shell of which is formed by 12 water molecules.

5 Conclusion

1. The equation of the ion oscillations in the solution for the case of "asymmetric" electric field is determinated. A mathematical model, which allows a better understanding of the process of separation of the solvated ions under the influence of an electric field and calculate required frequency for excitation of selective drift effect of cations of different chemical elements, is created.
2. Fundamental oscillation frequencies of the ion in the salt solution in polar dielectric fluid are within the interval from tens of Hz to units of kHz and correspond to two possible models of the solvated ion:
   a) the solvated shell mass does not exceed the value of 10^{-22} kg, the outer radius of the shell is not bigger than 0.05 μm, which is by an order of magnitude greater than the first solvation radius. When the solvent is water, about 10^4 water molecules are within the solvation shell. And the fundamental oscillations of the rotator will be tens of Hz;
   b) only relatively mobile molecules of the solvent which are not closely connected by the central ion electric field and border to the surface with the radius R_2 take part in the rotational motion. The value R_2 is defined from the condition $E_r(R_2) = E_{cr}$. The total

amount of such mobile molecules is significantly smaller than the number of the solvent molecules associated around the central ion. The mass of such a spherical rotator, formed by the layer of the solvent molecules fitting to the external border of the solvation shell, does not exceed 10^{-23} kg (350…400 water molecules). As this takes place, the fundamental frequencies of the rotator are units of kHz.

On the example of yttrium it is shown the possibility of the obtained results using as a basis for technology of rare earth elements separation in water solutions of their salts.

References

1. M.A. Kazaryan, I.V. Lomov, I.V. Shamanin, *Electrophysics of Structurized Saline Solutions in Fluid Polar Dielectrics* (Fizmatlit, Moscow, 2011)
2. R.F. Feynman, *Feynman Lectures on Physics: transl. from English* (Mir, Moscow, 1976)
3. A. Parcell, *Berkeley Course of Physics: transl. from English: In 5 Vol. — 3rd revised edition. Vol. 2: Electricity and Magnetism* (Nauka, Moscow, 1983)
4. M.A. Kazaryan, I.V. Shamanin, I.V. Lomov, Theoretical Basis of Chem. Technol. **44**, 1, 1 (2010)
5. E. Kamke, *A guide on ordinary differential equations: transl. from German* (5th edition, Nauka, Moscow, 1976)
6. I.A. Lukovskiy, M.A. Chernova, Akusticheskij Vestnik, **14**, 3, 23 (2011)

Surface Properties of Metal Hydroxide Microparticles in the Ambient Air

Valery Zakharenko[1,*], *Elena* Daibova[2], and *Natalia* Karakchieva[3]

[1]Boreskov Institute of Catalysis, Novosibirsk, Russia Federation
[2]Siberian Research Institute of Agriculture and Peat, Branch of the Siberian Federal Agrobiotechnology Research Center, Russian Academy of Sciences, Tomsk, Russia Federation
[3]Tomsk State University, Tomsk, Russian Federation

Abstract. The adsorption and photoadsorption properties of $Mg(OH)_2$ and $Ca(OH)_2$ microparticles in the ambient air were investigated. The compositional analysis of an adsorption layer of microparticles was carried out. The kinetics of photodesorption of molecules from microcrystal surfaces and the interaction of HCFC-22 (CHF_2Cl) in the dark and under light were studied. Quantum yields and their spectral dependencies were determined for CO_2 photodesorption, O_2 and CO photoadsorption. The effect of weakly bound CO displacement from the surface of microparticles was revealed during dark adsorption of HCFC-22. It is supposed that adsorbed CO is formed as a result of atmospheric CO_2 reduction after the break of Mg—OH bonds. In case of calcium hydroxide, CO is generated during the interaction of calcium hydroxide with carbon dioxide in the presence of water.

1 Introduction

From compounds containing a rare-earth element cerium, cerium dioxide CeO_2 is most widely used [1]. Introduction of cerium as addition in calcium and magnesium hydroxides substantially changes their physico-chemical properties. Before the study of these changes, it is necessary to investigate physico-chemical properties of alkaline-earth metals hydroxides.

This study is aimed at investigating of surface properties of magnesium hydroxide microparticles as well as calcium hydroxide under the conditions close to real atmosphere.

2 Experimental part

2.1 Calcium hydroxide

Industrial powdered calcium oxide ("pure" grade, "Reachem", Russia) was used as a starting compound for producing calcium hydroxide. In order to exclude in CaO the presence of calcium carbonate generated during the long-term storage of CaO, calcium oxide was heated in the air at 930 K for 4.5 hours. Due to this, calcium carbonate was decomposed into calcium oxide and carbon dioxide. The powder was cooled to room temperature (~300 K). Then a CaO suspension was prepared in distilled water and applied to the inner wall of a cylindrical quartz reactor. The surface area covered with 0.03 cm $Ca(OH)_2$ film was 70 cm^2. Calcium oxide was converted to calcium hydroxide $Ca(OH)_2$. Then water was pumped out from the reactor by a backing pump since water facilitated the conversion of calcium hydroxide to calcium carbonate in the air [2, 3]. In this way, a dense layer of calcium hydroxide without calcium carbonate impurities was supported on the reactor wall. Next, the reactor was soldered to a high-vacuum system for experiments. The reactor can be evacuated to 10^{-7} torr.

2.2 Magnesium hydroxide

For preparation of magnesium hydroxide sample, the $Mg(OH)_2$ brucite crystal has been used. The brucite crystal from the Tuva deposit was provided by the Geological Museum of the Institute of Geology and Mineralogy, Siberian Branch of the Russian Academy of Sciences, Russia. Under preparation of magnesium hydroxide sample the brucite crystal was ground in an corundum mortar in the air at room temperature. The specific surface for the obtained powder was ~ 2 m^2/g.

A suspension was prepared in distilled water from the powdered magnesium hydroxide sample by stirring and supported on the internal wall of the cylindrical quartz reactor. The reactor was similar to the reactor for $Ca(OH)_2$. Then the adsorption layer was dried in the air. The reactor with $Mg(OH)_2$ was soldered to the high-vacuum unit consisting of a gas supply system, a Pirani gauge and APDM-1 mass spectrometer (made in SU). The quantum yield of the photoadsorption (photodesorption) was found as the ratio of the quantity of photoadsorbed (photodesorbed) molecules to the number of quanta passed through the reactor's frontal (transparent) wall.

*Corresponding author: zakh@catalysis.ru

The methods for determining concentration and composition of gases in reaction mixtures, as well as the methods for calculating quantum yields and their spectral dependences are in detail described in [4, 5].

3 Results and discussion

3.1 Calcium hydroxide

In a number of conversions that occur under tropospheric conditions ($CaO \rightarrow Ca(OH)_2 \rightarrow CaCO_3$), the band gap of these compounds varies from 7.03 eV (calcium oxide) [6] to 7.3 ÷ 7.6 eV (calcium hydroxide) [7] and 5.0 eV (calcium carbonate) [8]. Due to high water vapor content (as compared to carbon dioxide), calcium oxide is converted to calcium hydroxide with the absorption spectrum shown in Fig.1 (curve 1). Its spectral dependence of absorption is close to that of calcium carbonate (Fig.1, curve 2) produced by dispersing the calcite crystal, if the absorption spectrum is registered in the shortest possible time after grinding the crystal. At the same time, if a cuvette with the powdered calcite sample is kept in the air for one month, the peak at a wavelength of 326 nm appears on the spectral dependence of absorption (Fig.1, curve 3). An inflection in spectral dependencies of absorption is recorded in calcium hydroxide (curve 1) and calcite (curve 2) in this spectral region. Any absorption in this spectral region is not observed for industrial calcium carbonate samples [9].

The study [9] supposes that the dispersion of the calcite crystal leads to the partial decomposition of calcium carbonate to CaO and CO_2. In its turn, CaO in the presence of water vapors in the air is converted to calcium hydroxide. This explains why spectral dependences of absorption of $Ca(OH)_2$ prepared in our experiment and $CaCO_3$ produced from the calcite crystal are close to each other.

After pumping out the reactor with the calcium hydroxide sample through a trap with a cool mixture (at 173 K) for a short time (~ 2 min), in reaction volume are revealed nitrogen oxides (NO and N_2O in the ratio of 1:1). These compounds are mostly desorbed from the surface of aerosol particles. It should be noted that carbon dioxide in the gas phase is not detected. The total number of molecules desorbed for a long time corresponds to 10^{17} molecules or 0.2% of the surface monolayer of the total sample in the reactor.

The mass spectrometric analysis of gases released from the surface of microcrystals showed that their primary component is nitrogen oxide (NO), with the total number of molecules in the reaction volume equal to $3 \cdot 10^{14}$ molecules.

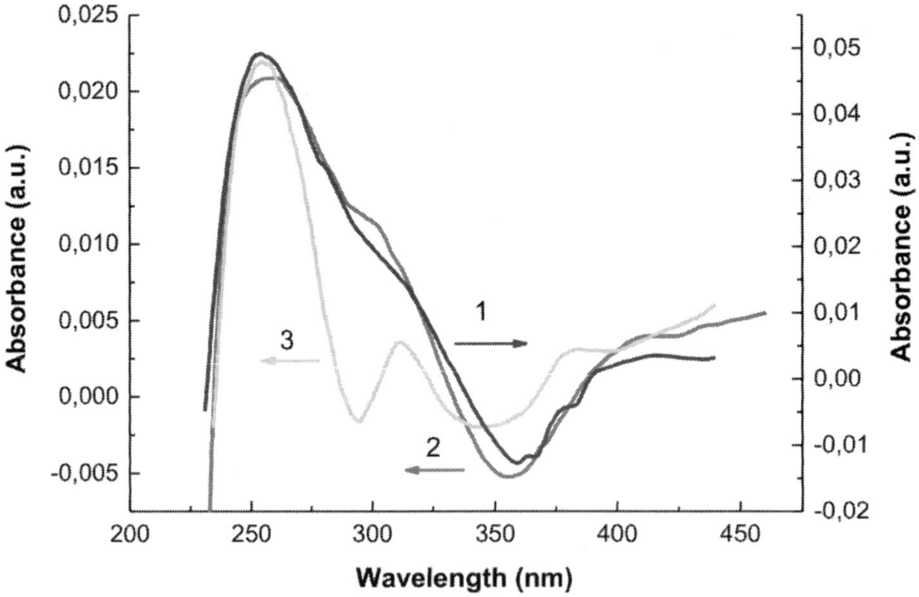

Fig. 1. Absorption spectra for the samples of calcium oxygen-containing compounds: (1) the layer of $Ca(OH)_2$ microparticles, (2) the layer of $CaCO_3$ microparticles obtained by dispersing the crystal of the mineral calcite, and (3) the second sample after one month keeping under ambient air.

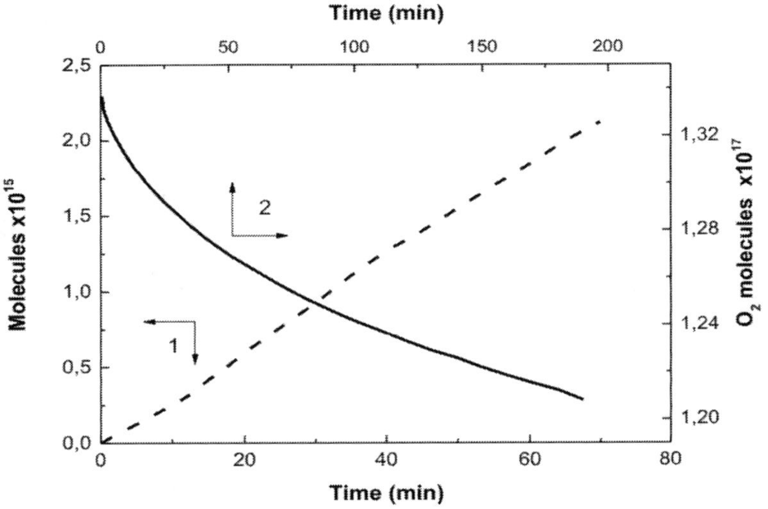

Fig. 2. Kinetics for the $Ca(OH)_2$sample: (1) photodesorption from the surface under UV filter and (2) "dark" adsorption of oxygen.

When the surface of $Ca(OH)_2$ particles is illuminated through an ultraviolet filter by the OSL-1 illuminator (made in SU) to the fullest extent, gases are photodesorbed from the sample surface. The photodesorption kinetics is shown in Fig. 2, Curve 1. According to the mass spectrometry data, nitrogen oxide NO was the primary gas in the reactor volume after photodesorption from the $Ca(OH)_2$ surface for 70 minutes.

When the NO acceptor molecule is photodesorbed, dark adsorption of the acceptor oxygen molecule takes place on adsorption centers of the surface that have been occupied by the NO acceptor molecule. The kinetics of this adsorption is shown in Fig.2, curve 2.

The numbers of adsorption centers that are occupied as a result of dark oxygen adsorption are an order of magnitude higher than the number of centers that become vacant after NO photodesorption. This indicates that oxygen adsorbed from the ambient air in the dark during the sample preparation frees surface acceptor centers, e.g., when reacting with water molecules of the reactor gas phase to produce hydrogen peroxide during the radiation of aerosol microparticles. This is confirmed by a significant increase in the seventeenth mass (OH^+) from trace amounts of hydrogen peroxide (HO-OH) in the mass spectrum as shown by the compositional analysis of gases in the reactor after photodesorption.

After dark adsorption of oxygen, the initial rate of oxygen photoadsorption was measured at a pressure of 10^{-2} torr. The rate was equal to $6 \cdot 10^{13}$ molecules·sec^{-1} when irradiated through the UV filter. For comparison, the initial photodesorption rate was 20 times lower under the same illumination conditions.

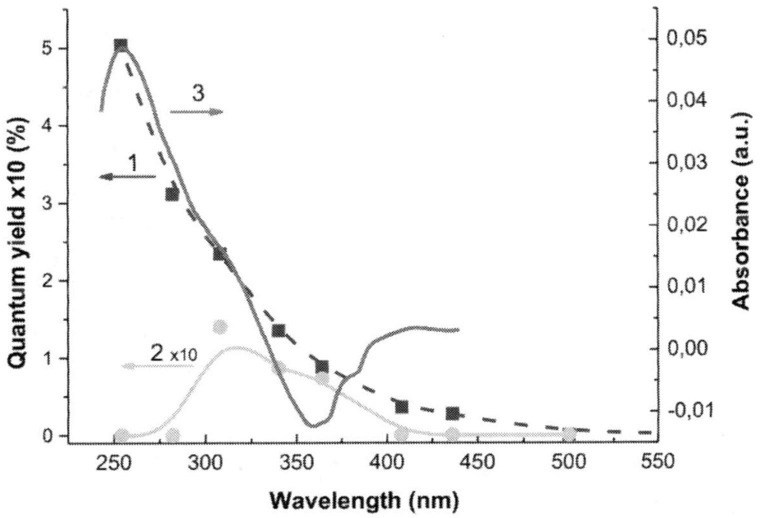

Fig. 3. Spectral dependencies for the $Ca(OH)_2$ sample: (1) quantum yield of O_2 photoadsorption, (2) quantum yield of photodesorption (scale up 10 times), and (3) absorbance relative to a reference standard (powdered MgO).

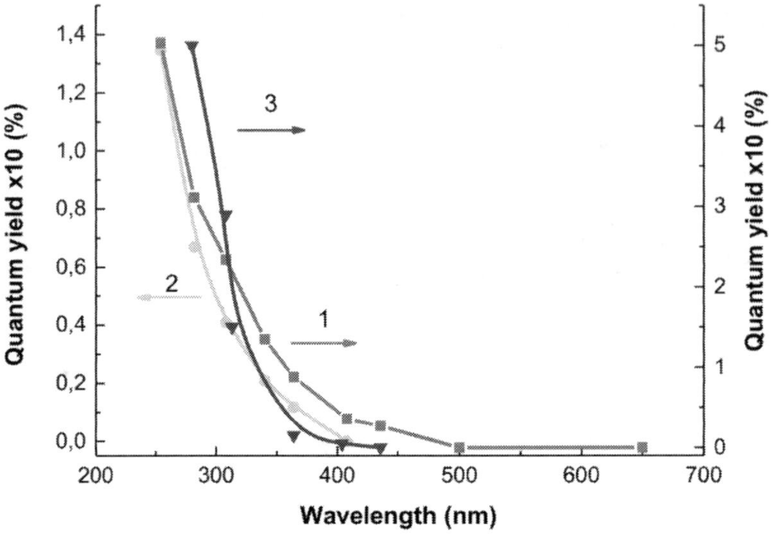

Fig. 4. Spectral dependencies of quantum yield of O_2 photosorption: (1) $Ca(OH)_2$, (2) $CaCO_3$ [9], and (3) CaO [10].

Oxygen photoadsorption and photodesorption for the calcium hydroxide sample have also different spectral dependences of quantum yields (Fig.3, curves 1 and 2 for photoadsorption and photodesorption, respectively).

Photodesorption from the surface of $Ca(OH)_2$ microparticles is observed in the spectral range from 275 nm to 400 nm (Fig. 3) with the maximum of around 300 nm, i.e. in the same spectral range as photodesorption from the surface of the sample produced by dispersing the calcite crystal ($CaCO_3$) [9].

In its turn, the spectral dependence of oxygen photoadsorption for the sample tested in this study is close to that of the calcite sample as shown in Fig.4 (curves 1 and 2 for $Ca(OH)_2$ and $CaCO_3$, respectively). Furthermore, according to [9], photoadsorption of oxygen on calcium oxide sample after its high-temperature vacuum treatment is also observed at wavelengths of shorter than 400 nm (Fig. 4, Curve 3).

According to [10], photoadsorption of oxygen in CaO occurs after absorption of a photon and electron transfer to the surface level (surface absorption of calcium oxide). Oxygen of the gas phase forms adsorbed oxygen $(O_2^-)_s$ with this electron. Both for dark oxygen adsorption in calcium oxide [11, 12] and oxygen photoadsorption in calcium oxide [10], anion vacancies of the CaO — oxygen crystal structure must be present. In this case, electron transfer from an anion vacancy that occurs as a result of photon absorption leads to the subsequent oxygen adsorption.

It can be assumed that spectral dependencies of quantum yield of oxygen photoadsorption for calcium hydroxide (Fig. 4, curve 1) and calcium carbonate (Fig.4, curve 2) are close to that for calcium oxide (Fig.4, curve 3) due to the formation of close (in terms of energy) anion vacancies in oxide, hydroxide and calcium carbonate lattices. Vacancies are generated as a result of decomposition of $Ca(OH)_2$ exposed to irradiation, as well as when the $CaCO_3$ crystal is dispersed to CaO, while CaO is reduced by radiating it in the presence of water as a reducing agent.

3.2 Magnesium hydroxide

Experiments with magnesium hydroxide samples started after degassing the reactor volume with $Mg(OH)_2$ at room temperature through a trap cooled to 173 K. CO_2, NO and CO were observed in desorption products. After 20 minutes of the accumulation, the ratio between volumes of these gases was equal to 10:0.1:1, respectively, for CO_2, NO and CO. In case of long-term accumulation of desorption products for 24 hours, the ratio changed to 20:10:1. The kinetics of carbon monoxide accumulation had a characteristic time of several minutes, while the partial pressure of desorbed CO in the gas phase reached an equilibrium value. This may indicate a weak interaction of CO with the surface of magnesium hydroxide. Slower kinetics of NO desorption (with the characteristic time of hundreds of minutes) suggest that it is formed as a result of the slowly reconstructed $Mg(OH)_2$ surface generated during the destruction of the brucite crystal in the air. Any gases (other than CO and NO) not frozen out at the liquid nitrogen temperature (77 K) were not identified in products of the gas released from the surface of magnesium hydroxide.

Desorption of nitric oxide was observed also for TiO_2 and MgO samples produced by grinding rutile and periclase crystals, respectively, in the air [4, 5].

In addition to water molecules, the surface (0001) produced from the destructed $Mg(OH)_2$ also absorbs, in the dark at room temperature, molecules of carbon dioxide and nitrous oxide contained in the air in considerable quantities [13]. The number of adsorbed molecules of these gases depends on partial pressure of the gas in the atmosphere and the adsorption capacity of the magnesium hydroxide surface in relation to these gases.

When dispersing brucite, periclase and rutile crystals, generated surfaces of microcrystalline aerosol particles have different structures of cleavage planes {0001}, {001} and {110}, respectively, for $Mg(OH)_2$ [1], MgO [14] and TiO_2 [15]. They also differ in terms of chemical nature of surface atoms that remain after break of the respective bonds: Mg-OH for $Mg(OH)_2$, Mg-O for MgO and Ti-O for TiO_2. However, the ratio of nitrogen monoxide to carbon monoxide is similar for these three powdered samples.

Moreover, desorption of gases from the particle surface of these three aerosol samples is characterized by common features, namely:

1. carbon monoxide is weakly bound to the microparticle surface generated during the dispersion of a single crystal in the air;
2. nitrogen monoxide is slowly released from the surface as a result of its reconstruction;
3. carbon dioxide is the primary product desorbed from the surface into the gas phase.

While carbon dioxide may arise from desorption of atmospheric carbon dioxide [13], adsorbed CO and NO are likely to be result of the interaction between gas molecules of the atmosphere with the surface generated during the dispersion of the crystal in the air. For example, carbon monoxide may arise from the reduction of carbon dioxide (CO_2 content in the air is about 340 million^{-1}) by a surface metal atom with an unsaturated bond:

$$(Me—)_s + CO_2 \rightarrow (CO)_s + (MeO)_s \quad (1)$$

Nitrogen monoxide is formed by oxidation of N_2O with oxygen followed by decomposition of N_2O_2 dimer to nitrogen monoxide:

$$(O—)_s \; + N_2O \rightarrow [Vac]_s + (N_2O_2)_s \quad (2)$$
$$(N_2O_2)_s \rightarrow 2\,(NO)_s \quad (3)$$

N_2O is adsorbed from the gas phase of the atmosphere during the sample preparation (0.25–0.35 million^{-1} of N_2O as a component of the uncontaminated air).

Illumination of the $Mg(OH)_2$ surface through the UV filter leads to additional desorption, mainly carbon dioxide and small amount of carbon monoxide (in the ratio of 1:10). When illuminated through the UV filter, the photodesorption rate of carbon dioxide was equal to $1.1x10^{13}$ molecules·sec^{-1}. The total number of photodesorbed molecules reached $1.2x10^{16}$ and corresponded to 0.1% of the surface monolayer of magnesium hydroxide. Photodesorption was irreversible, i.e. reverse dark adsorption did not occur after switching off the light.

After calculating the quantum yield of carbon dioxide photodesorption for different monochromatic radiation wavelengths, the spectral dependence of the quantum yield was determined. It is shown in Fig. 5 (Curve 1). The same figure shows the spectral dependence of the diffuse reflection of powdered magnesium hydroxide measured relative to magnesium oxide (Curve 2).

According to these results, the photodesorption activity of $Mg(OH)_2$ in relation to carbon dioxide is observed in the same region of the spectrum in which light absorption by the adsorbed layer of magnesium hydroxide (wavelengths less than 400 nm) is recorded.

In contrast to irreversible photodesorption of CO_2, when the $Mg(OH)_2$ surface is illuminated through the UV filter in the presence of CO, its reversible photodesorption is observed at the initial rate of $4.2x10^{13}$ molecules · sec^{-1}. The amount of photoadsorbed CO reached $7.1x10^{13}$ molecules · sec^{-1}. After turning off the light, the photoadsorbed CO is completely released into the gas phase.

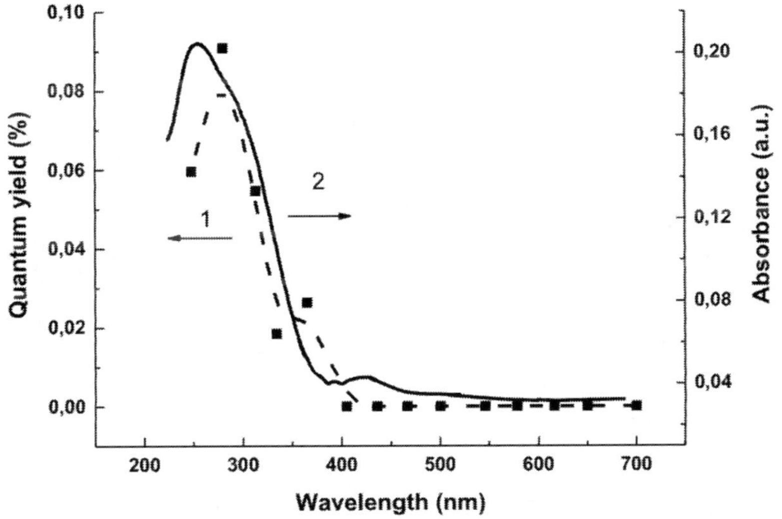

Fig. 5. Spectral dependencies for the $Mg(OH)_2$ sample: (1) quantum yield of CO_2 photodesorption on the microparticles obtained by dispersing the crystal of the mineral brucite and (2) absorbance relative to a reference standard (powdered MgO).

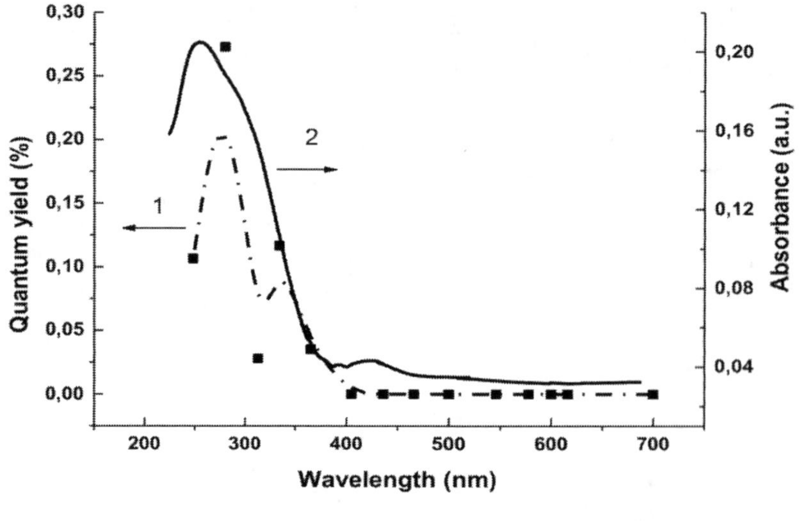

Fig. 6. Spectral dependencies for the $Mg(OH)_2$ sample: (1) quantum yield of CO photoadsorption on the microparticles obtained by dispersing the crystal of the mineral brucite and (2) absorbance relative to a reference standard (powdered MgO).

The spectral dependence of the quantum yield of CO photoadsorption (Fig. 6, Curve 1) is close to the spectral dependence of CO_2 photodesorption (Fig. 5, Curve 1) and to the spectrum of absorption by the adsorbed layer of magnesium hydroxide (Fig. 6, Curve 2).

Thus, the absorption of photons from a wavelength region of less than 400 nm leads to both photodesorption of CO_2 and photoadsorption of CO; however, the formation of CO_2 on the magnesium hydroxide surface is not associated with CO photoadsorption. The following mechanism of photoprocesses can be proposed:

$$V.B. + [Vac]_s \xrightarrow{h\nu} h_{v.b.} + [e]_s \quad (4)$$

$$(CO_2^-)_s + h_{v.b.} \rightarrow CO_2 \quad (5)$$

$$CO + [e]_s \rightarrow CO\text{—}[e]_s \quad (6)$$

where **V.B.** is the valence band of magnesium hydroxide, $\mathbf{[Vac]_s}$ is the surface vacancy, $\mathbf{h_{v.b.}}$ is the "free" hole in the valence band, $\mathbf{[e]_s}$ is the electron in the valence band, $\mathbf{CO\text{—}[e]_s}$ is the weakly bound carbon monoxide.

3.3 Interaction of HCFC-22 with the $Mg(OH)_2$ and $Ca(OH)_2$ microparticle surfaces

During the adsorption of HCFC-22, its dark adsorption rate exceeds the rate of simultaneous CO desorption from the $Mg(OH)_2$ surface. While in case of industrial MgO, the dark adsorption rate of HCFC-22 is 4 times less than the rate of simultaneous CO desorption [16], in case of $Mg(OH)_2$ the adsorption rate of HCFC-22 is two times higher than the CO desorption rate.

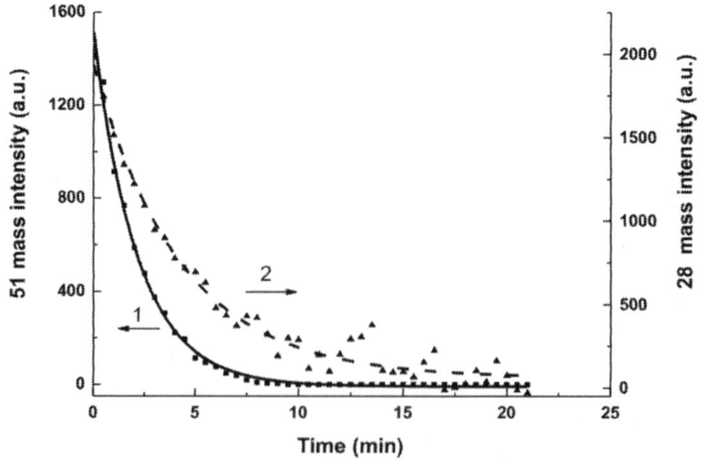

Fig. 7. Kinetics for the $Mg(OH)_2$ sample: (1) "dark" adsorption of Freon 22 (51 mass intensity) and (2) decreasing the adsorbed CO amount from the surface (28 mass intensity).

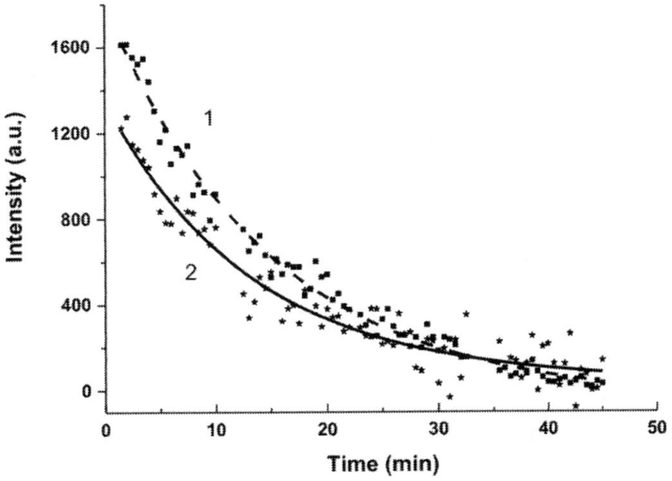

Fig. 8. Kinetics for the $Ca(OH)_2$ sample: (1) "dark" adsorption of Freon 22 (51mass intensity) and (2) decreasing the adsorbed CO amount from the surface (28 mass intensity).

According to the mass spectrometry data for the second supply of HCFC-22, both its adsorption kinetics and CO desorption kinetics are approximated by first-order equations (Fig. 7, Curves 1 and 2, respectively, for HCFC-22 and CO). With an increasing amount of HCFC-22 adsorbed in the dark, its adsorption rate decreases after pumping out produced carbon monoxide and supplying a "new" portion of HCFC-22.

At the first supply of HCFC-22, the dark adsorption rate of HCFC-22 on the surface of a part of calcium hydroxide is close to the rate of carbon dioxide release from the $Ca(OH)_2$ surface. Further, when carbon monoxide is pumped out and a "new" portion of HCFC-22 is supplied, the dark adsorption rate of HCFC-22 decreases but still exceeds the CO desorption rate (Fig. 8).

Thus, adsorption of HCFC-22 displaces the weakly adsorbed carbon monoxide, a part of which is desorbed when $Mg(OH)_2$ and $Ca(OH)_2$ samples are pumped out. Magnesium hydroxide is known to have high adsorption capacity relative to chlorine compounds and is used as a component of mixtures for removing such compounds from gases [17]. The high adsorption capacity of magnesium and calcium hydroxides is indicated by a high degree of filling $Mg(OH)_2$ and $Ca(OH)_2$ microparticle surfaces observed in this study: it exceeds 10% of their surface monolayer.

In our opinion, the formation of weakly bound carbon dioxide on the surface of MgO and $Mg(OH)_2$ produced by dispersing MgO periclase and $Mg(OH)_2$ brucite crystals is associated with broken Mg – O [5] and Mg – OH bonds (Equation 1) and the reduction of adsorbed carbon dioxide to carbon oxide. However, the calcium hydroxide sample was not produced from the $Ca(OH)_2$ portlandite crystal. In this case, it can be assumed that the formation of carbon oxide in the reaction volume of adsorbed water and water in gas phase takes place during the interaction between $Ca(OH)_2$ and carbonic acid, e.g., through the reaction:

$$Ca(OH)_2 + H_2CO_3 \rightarrow CO + H_2O_2 + CaO + H_2O \quad (7)$$

After having been saturated with HCFC-22, surfaces of magnesium and calcium hydroxides samples were exposed to light through the UV filter which transmitted radiation with wavelengths of 235 to 410 nm. The mass spectrometry data show that, when exposed to such light, photoadsorption of HCFC-22 and, at the same time, photodesorption of carbon monoxide are observed. Using the Pirani gauge, we recorded a decrease in the total pressure at the rate of ~ 10^{12} molecules · sec $^{-1}$.

Thus, the surface of $Mg(OH)_2$ and $Ca(OH)_2$ aerosol microparticles exposed to light in the presence of HCFC-22 results in a slight increase in its adsorption rate as compared to dark adsorption of HCFC-22.

4 Conclusions

Thus, the study of calcium hydroxide without calcium carbonate impurities demonstrates that there is no carbon dioxide on the surface of the calcium hydroxide aerosol microparticles. The photochemical activity in relation to oxygen photoadsorption and photodesorption of gases from the surface of aerosol particles is observed at the wavelengths of photons shorter than 400 nm. Photoadsorption of oxygen in oxygen-containing compounds of CaO, $Ca(OH)_2$ and $CaCO_3$ have similar mechanisms associated with the formation of anion vacancies in the lattice structure of these compounds.

The effect of weakly bound CO displacement from the surface of magnesium and calcium microparticles was revealed during the dark adsorption of HCFC-22. It is supposed that carbon monoxide is formed as a result of atmospheric CO_2 reduction after breaking Mg—OH bonds during the dispersion of the brucite crystal in the air. The formation of carbon oxide on the calcium hydroxide surface in the reaction volume of adsorbed water and water in gas phase takes place during the interaction between $Ca(OH)_2$ and carbonic acid.

This work was conducted within the framework of budget project No. 0303-2016-0002 for Boreskov Institute of Catalysis.

References

1. K. Reinhardt and H. Winkler in "Cerium Mischmetal, Cerium Alloys, and Cerium Compounds" in *Ullmann's Encyclopedia of Industrial Chemistry* (Wiley-VCH, Weinheim, 2000)
2. Ch. Theocharis, D. Yeates, Colloids and Surface, **58**, 353 (1991)
3. G. Montes-Hernandez, A. Pommerol, F. Renard, P. Beck, E. Quirico, O. Brissaud, J. Chem. Eng., **161**, 250 (2010)
4. V.S. Zakharenko and E.B. Daibova, Atmos. Ocean Optics, **22**, 611 (2009)
5. V.S. Zakharenko and E.B. Daibova, Atmos. Ocean Optics, **24**, 516 (2009)
6. R.C. White, W.C.Walker, Phys. Rev. Lett., **22**, 1428 (1969)
7. A. Pishtshev, S.Zh. Karazhanov, M. Klepov, Computation Mater. Sci., **95**, 693 (2014)
8. M.G. Brik, Physica B: Condensed Matter., **406**, 1004 (2011)
9. V.S. Zakharenko and E.B. Daibova, Atmos. Ocean Optics, **28**, 540 (2015)
10. A.M. Volodin, V.A. Bolshov, T.A. Konovalova, Molecular Eng.,**4**, 221 (1994)
11. D. Cordishi, V. Indovina, M.Occhiuzzi, J. Chem. Soc. Faraday Trans. I, **74**, 883 (1978)
12. M. Che, S. Coluccia, A.Zecchina, J. Chem. Soc. Faraday Trans. I, **74**, 1324 (1978)
13. P. Brimblecombe, *Air Composition and Chemistry* (Academic Press, New York, 1988)
14. S. Benedetti, P. Torelli, P. Luches, E. Gualtieri, A. Rota, S. Valeri, Surf. Sci., **601**, 2636 (2007)
15. W.S. Epling, Ch.H.F. Pedan, M.A. Henderson, U. Diebold, Surf. Sci., **412/413**, 333 (1998)
16. V.S. Zakharenko and V.N. Parmon, Kinet. I Katal., **41**, 834 (2000)
17. X.-F. Wu, G.-H. Hu, B.-B. Wang, Y.-F. Yang, J. Crystal Growth., **310**, 457 (2008)

Data Visualization Tools for Materials Properties Research

Artem Obkhodsky[1,*], *Sergey* Kuznetsov[2], *Alexander* Popov[1], *Aleksey* Zolotarev[2], and *Victor* Sachkov[3]

[1] National Research Tomsk Polytechnic University, Tomsk, Russian Federation
[2] LLC "Depos", Tomsk, Russian Federation
[3] National Research Tomsk State University, Tomsk, Russian Federation

Abstract. In work presented the results of research the algorithms of experimental data visualization, which obtain in the simulation of materials. The algorithms study was carried out on the example of the complete cycle for calculating the properties of an ideal diamond crystal. The calculation of crystal lattice properties was carried out using a software package for materials simulation. The obtained experimental data were analyzed using program REM_Viewer in which have been realized the algorithms of data visualization.

1 Introduction

Modern graphic systems ensure satisfactory performance when designing complex dynamically animated images [1]. Computer graphics allows demonstrating physical processes and this contributes to their better understanding [2]. It is a well-known fact that represented information (animation, graphics) can be perceived more effectively [3, 4]. Conversely, the crystalline structure and different spatial physical distribution values obtained via computer simulation methods are more understandable in a 3D interpretation [5].

Presently, computer simulation of materials provides a high predictive capacity and significant potential in developing new materials with predefined properties and characteristics. [6, 7]. Computer simulation is based on valid multilevel physical models of material structure and properties and multi-scale computing approach by applying quantum chemistry methods and classical methods of molecular dynamics. [8].

In recent years it activated interest of researchers in the modeling and prediction of crystal structures properties. This interest is due to a number of fundamental and applied problems in the solution which can be used prediction algorithms. For example, when studying phenomena such as crystalline polymorphism [9], the change of solid materials' phase state under various conditions [10]. To applied problems can be attributed the search of new drug compounds with polymorphous phase [11], high-density energy materials [12], the new nonlinear optical materials [13].

Functions of materials structure visualization on atomic-level are particularly important when comparing the results of numerical experiments obtained by software package for materials simulation with the results of experimental studies, and for verification of mathematical models and methods used in the software package [14, 15]. Besides, the graphic image of material's atomic-molecular structure allows identify the most significant zones of materials restructuring during the physical and chemical processes.

The results of quantum chemistry calculation are the data files of extensive numerical information. To operate data bulk contained in output file used special program-interpreters for quantum chemistry calculation results. These programs with 3D graphics provide access to calculation results and their preliminary analysis [16].

From the existing solutions for data visualization in the area of materials modeling should note Diamond, Topos, SHELXL97, CRYSTALS and PLATON [17]. Although each program exhibits individualization, in some cases, they exhibit either an extended functionality or even narrow specificity, which, in its turn, excludes their application in 3D visualization of crystalline structure and different spatial physical distribution values.

Accepting the abovementioned into consideration, the aim of the work was to create algorithms and data visualization tools for materials properties research which provides 3D dynamical visualization of crystalline structure and spatial distribution of materials physical parameters. Research of functions and characteristics of data visualization algorithms and a program were carried out on the example of numerical calculation the properties of an ideal diamond crystal.

2 Data visualisation algorithms

REM_Viewer is based on algorithmic programming language C++ and is implemented into computing REM-based material modeling software package, functioning on grid computing basis. This program could be used in free-running mode.

REM_Viewer software package includes 3D visualization of material structure, configurable graphic development with visualization domain customization.

* Corresponding author: art707@tpu.ru

Intercommunicating with process directory project REM_Viewer forms source data for the experiment, forms and reads transport UTF-file data, reads and processes arrays containing information on material structure and properties. The module complex includes other intercommunicating components.

To operate the front end application the following components are included: ECM with x86, x64 architecture and processor Intel Core i3-4130 2x3.2 GHz or better; RAM capacity- not less than 2 gilbert; ROM capacity- not less than 20 gilbert; video display adapter with supporting display image resolution of not less than 1280×1024 pixel and supportive OpenGL 2.0 or better; installed OS Linux with core version not less than 2.6.20 and/or OC – MS Windows XP/Vista/7 (32 bit, 64 bit); installed Driver OpenGL 2.0 and more; as well as display- video display adapter compatible.

Figure 1 illustrates the program module diagram within REM_Viewer.

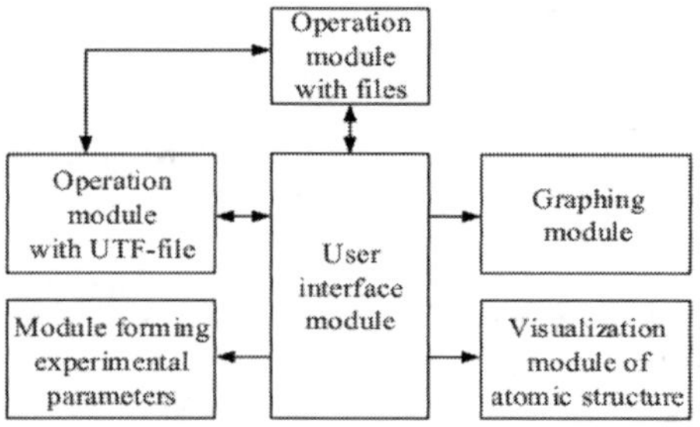

Figure 1. Diagram of intercommunicating internal REM_Viewer modules

REM_Viewer start-up involves creating experiment process directory path and project file fault-based testing. This file is applied to cast to the user interface program at the moment of its close-down.

Forming a new experiment starts by inputting basic parameters. This launches the formation of a new UTF-file and its progressive filling. Basic parameters include work directory path, experiment description and date; researcher name, surname and e-mail; and institution name.

Each experimental parameter has its own identifier and level. All experiment parameters are displayed on main program window (Fig.2), in the list “Materials’ parameters”. All parameters are divided into two groups: initial and calculated. Calculated parameters are automatically updated and added throughout the experiment. On the left-side of the window all files of current experiment and relevant files (algorithm calculation executables, baseline data set files, visualization files, LOG-files, etc.) are output. All initial, calculated and other experimental parameters as well as process directory structure are recorded in UTF-file on demand.

At the bottom REM_Viewer main window event output area is located. In the course of program operation the following event output appears describing the user response. All messages in occurrence of events are divided into informational and warning error messages and fatal error messages In this case, error message coloring is applied and then stored in archive log as LOG-file.

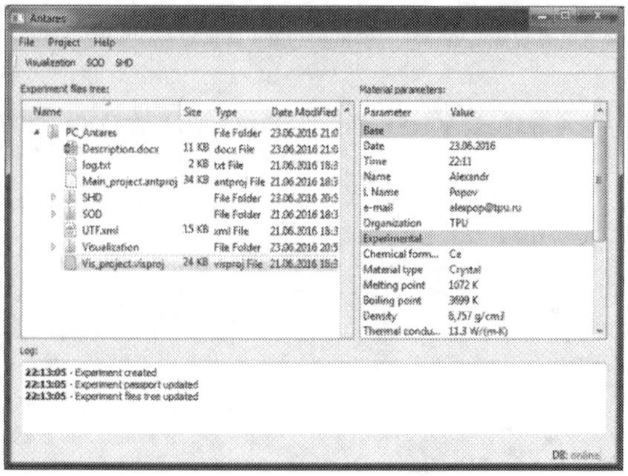

Fig. 2. Program REM_Viewer main window

The key aspect of this research is the analysis of electron density distribution under the following given conditions: nucleus pattern, atom coordinates in a lattice and their interaction. This information is in the output files available from quantum- chemical calculations and represents an unstructured block of given data format. Visualization results and all configuration tools are found in the window as shown in the Fig. 3. Conducting experiments in materials simulation specific output parameters are formed, such as: material property as independent numerical values, dependence as value series, static allocation, time-dependent allocation, 3D-forms, describing molecule structure and/or crystal cell. REM_Viewer is used for visualization of all above-mentioned parameters. To provide easily available reviewing information one can select displayed parameter list in front end application.

Graphic data visualization in REM_Viewer carried out across interacting with function library OpenGL.

Compatibility of data format and protocols poses a problem in visualizing and distributing system design [19]. Depending on presented input data REM_Viewer allows to build two graph types: static and dynamic. Data for static graphing is one-dimensional array. This graphic type shows the single one-dimensional array related to another. Plotting static graphics is possible by two- or three-axes, which could have different scales. If the second variant is chosen, then specified parameters would embrace common axis X. Several data can be chosen for array axes Y1 and Y2.

The second graph type is dynamic. In this case graph data includes two-dimension array, each column of which corresponds to one-dimension array value. This graphic is applied if it is necessary to reflect the process dynamics. For example, the experiment showing temperature distribution on work surface, i.e. temperature value is measured several times within specified time intervals. Then each temperature array value depending on distance will correspond to a specific time interval. Thus, moving the thumb time in the window to visualization result, re-plotting will correspond to the data relevant to this time interval.

DOI: 10.1051/matecconf/20179600014

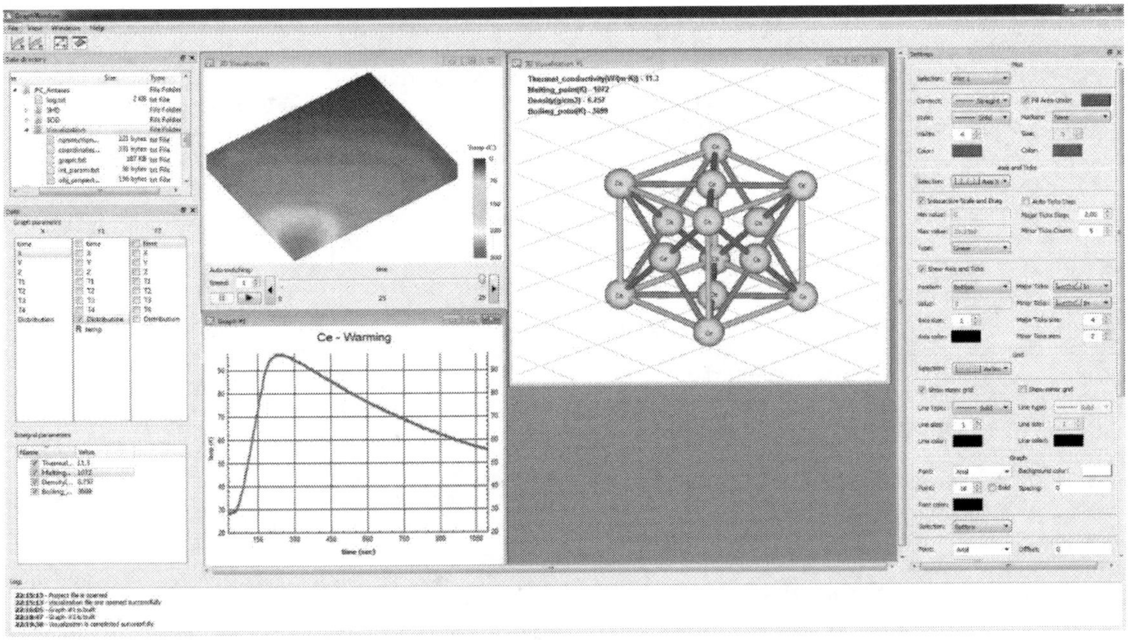

Fig. 3. REM_Viewer main window

Any number of graphs with different parameters and display settings can be developed in REM_Viewer. The user can change the graph scale in respect to any parameter, location and calculation module, represented coordinate axis and grid, types of graph lines, grids and axes, captions and legends, etc. All configurations are applied to current live graph, consequently, it is possible to chose different configurations and image parameters for various graphs. When switching from one graph to another visual setting up field updates in accordance to selected graph.

Besides, plotting graphs REM_Viewer embraces 3D visualization of the material structure. In this case text data is converted into interactive 3D image. Data for this visualization includes atomic coordinates set, determining their position in space, as well as information on atom bonds.

All atoms and their bonds could be divided into classes. One class includes atoms and bonds having similar properties. REM_Viewer reflects possible property changes of the atoms and bonds, as well as mode changes of these atoms and bonds, their size and color within any class. As a result the user could set up the visualization if necessary. Alternatively, possible changes of the output area mode could be planned. The user can also set up background color, coordinate axes and grids, as well as typeface and output position of material property parameters at one's own discretion. By analogy with graphs, it is possible to create several windows of visualized material structure and apply different configurations and visualization parameters within REM_Viewer.

Visualization algorithms of heat transfer and deformation processes (for example, temperature distribution for modeling materials) were implemented into REM_Viewer. Based on this visualization type one can analyze material impact resistance at any given point. In terms of data format visualization could be either static or dynamic.

To start visualization it is necessary to predetermine the following calculation parameters: exterior mold dimensions, initial molding temperature, grid simulation, experiment time, melting mold material temperature, specific thermal capacity and thermal conductivity of material, heater type and temperature, given point of material impact resistance and impact resistance time throughout the experiment. Results of quantum chemical calculation and reference data could be used as input parameters of the material. Corresponding to the predetermined parameters, the mold is divided into single cells forming the grid. The less the predetermined cell size, the more precise the visualization results and longer the calculation time of the whole mold. Determined input data is stored in a file and is passed to the software package for materials simulation. Based on this, the calculation for the dynamic pattern of heat transfer processes is performed. After calculating the data are stored for further visualization. Received total file includes the coordinates of all mold cells, corresponding to temperature values of material cells.

Color-coded method is used to reflect the temperature of each cell, i.e. cell temperature values are recorded as a palette, where, red corresponds to maximum temperature; blue - minimum temperature. After cell layout, each cell is assigned a color with respect to the coordinates. Every color is appropriate to the current palette. Interpolation of temperature values is implemented at cell interfaces for easy perception of visualization results.

Visualization in REM_Viewer could be either static or dynamic.

Static visualization is used before the complete of the

experiment calculation time interval. Receiving data from external calculation module could be either on-demand or after calculation completion. The time frame of numerical experiment and time-interval of data load must be specified before start visualization. This visually reflects the dynamic pattern of temperature distribution in the mold even before final calculations. Thus, this excludes the necessity to wait for the complete calculation of the time interval. This method should be used only if there is no confidence in the precision of given experiment parameters.

Dynamic visualization enables to observe the complete process of temperature distribution and/or material deformation throughout the experiment. The calculated module files include the data set for each experiment time step. In this case, visualization is implemented only after retrieving the result file. This is a drawback for this type of visualization as the calculation module of experimental data processing could be time-consuming. If moving the time slider, the temperature distribution pattern in the visualization window will be updated according to selected time-stamp.

3 Experiment

Experiment process was conducted via REV_Viewer by algorithm illustrated in the Fig.4.

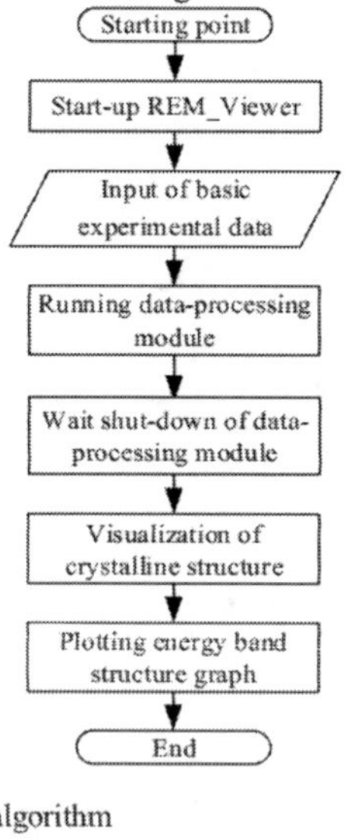

Fi. 4. Experiment algorithm

To test the functions of designed algorithms and a program was conducted the experiment on imaging data, obtained during simulating an ideal diamond crystal. software package for materials simulation. Calculation of crystal properties and characteristics of the material as a whole was carried out using a computational module which is along with the REM_Viewer included in the software package for materials simulation [20]. This module involved calculating the ideal diamond crystal coordinates by molecular dynamics method (MDM) at given temperature 295K and zero pressure. Further, calculations of energy band structure by Hartree-Fock [21, 22] method in an approximation to CO LCAO (crystal orbitals as linear combination of atomic orbitals) by applying cyclic cluster model were conducted to obtain the coordinates.

The experiment target was to obtain the geometrical characteristics of diamond crystal and the calculation of its energy band structure [23]. Based on all obtained calculation results the visualization of these experimental results was set up in REM_Viewer.

4 Results and Discussion

After the optimization of the crystalline lattice by computational module, the calculation of diamond crystal via Hartree-Fock method in an approximation to CO LCAO was conducted. Further the visualization of obtained crystalline lattice atom coordinate values was conducted. It was established that diamond crystalline lattice is face-centered cubic (space group – Fd3m) [24]. Visualization result is illustrated in the Fig. 5.

To calculate energy band structure the CO LCAO (crystal orbitals as linear combination of atomic orbitals by applying cyclic -cluster model) was applied. The graph was plotted based on these results and illustrated in the Fig. 6.

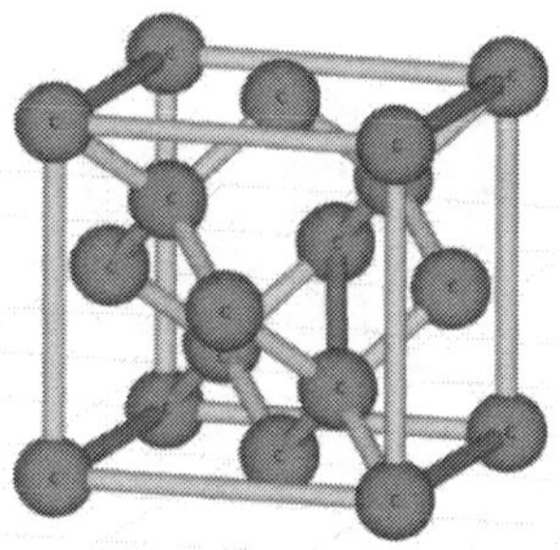

Fig. 5. Visualization result of diamond unit cell

Calculation results showed good agreement with the experimental data (ratio error is 3.78 %), which testify that the method parameters were correct.

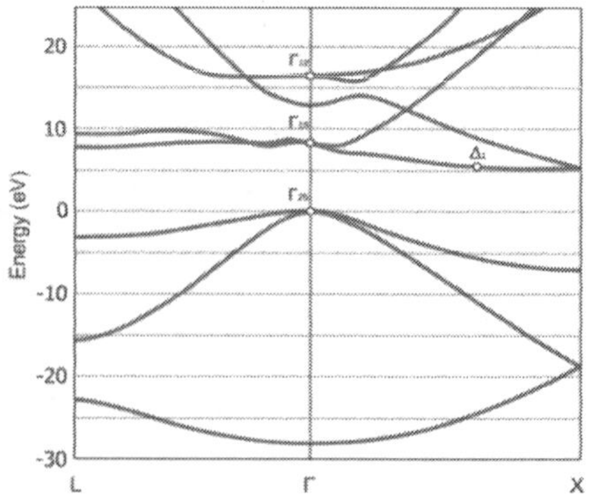

Fig. 6. Diamond energy band structure

Quantum-chemical calculation results are depicted below. Calculation method applied in this paper is denoted as CO LCAO CM (crystal orbitals as linear combination of atomic orbitals via cyclic-cluster method).

As a result, the following values were obtained:

- 0.76 eV (Δ minimum);
- 5.48 eV (Γ_{25} - Δ_1);
- 7.3 eV ($\Gamma_{25} - \Gamma_{15}$);
- 16 eV ($\Gamma_{25} - \Gamma_{12}$).

Coordinate calculations were implemented using optimized molecular dynamic method which introduced an error in the calculations. However, the results are in agreement with the experimental results but not including maximum valence band which is within the experimental data range.

It should also be noted that in this case the calculation involves ideal diamond crystal, consequently, no natural crystalline lattice defect is available. However, the calculation results are in good agreement with the experimental data.

5 Conclusion

As a result the algorithms of data visualisation and program REM_Viewer were designed. They provides graphic user interface of the software package for materials simulation. This allows to form initial data for experiments in calculating materials properties and to provide visualization file-stored experiment data.

The results of experiments on the calculation the properties of an ideal diamond crystal were confirmed by REM_Viewer, efficiency of which allows using it in further experiments in materials simulation. Further it is planned to implement into REM_Viewer algorithms which will provide visualisation the 3D distribution of physical values in space.

The work was performed with financial support of the Ministry of Education and Science of the Russian Federation. Grant agreement No. RFMEFI57814X0095, dated November 28, 2014.

References

1. M. Rofouei, T. Stathopoulos, S. Ryffel, W. Kaiser, M. Sarrafzadeh, *Proc. of the Conf. "Power Aware Computing and Systems"* (USENIX Association, San Diego, 2008)
2. V. Averbuh, *Proc. of the 1rst Int. Conf. "Three-Dimensional Visualization of Scientific, Technical and Social Reality. Cluster Modeling Technology"*, (UdGU, Izhevsk, 2009)
3. V. Borisov, A. Ignatenko, Computer Graphics and Multimedia (in Russian – Komputernaja Grafika i Multimedia), **4**, 3 (2006)
4. M. Friendly, *Department of Mathematics and Statistics* (York University, York, 2009)
5. V.A. Blatov, *Commission on Crystallographic Computing* (IUCr Newsletter, Chester, 2006)
6. I. Simonovski, L. Cizelj, Grain scale modelling Approach for polycrystalline agregates in *Polycrystalline Materials - Theoretical and Practical Aspects*, edited by Edited by Z. T. Zachariev (InTech, Rijeka, 2012)
7. A.Ju. Kuksin, A.V. Lankin, I.V. Morozov, Software Systems: Theory and Applications (in Russian – Programmnye Sistemy: Teorija i Prilozhenija) **1**, 191 (2014)
8. V.Yu. Popova, A.N. Xlystov, A.V. Bondin, Computer Studies and Modeling (in Russian – Komp'juternye Issledovanija i Modelirovanie), **1**, 137 (2016)
9. Dzh. Bernshtejn, *Polymorphism of Molecular Crystals* (in Russian – Polimorfizm molekuljarnyh kristallov) (Mir, Nauka, 2007)
10. I.I. Voroncov, K.A. Potehin, M.Ju. Antipin, Ja.Z. Voloshin, A.I. Stash, V.K. Vel'skij, I.I. Dubovik, B.C. Papkov, Crystallography (in Russian – Kristallografija), **46**, 833 (2001)
11. S.L Price, Adv. Drug Deliver Rev. **56**, 301 (2004)
12. A.V. Dzyabchenko, T.S. Pivina, E.A. Arnautova J. Mol. Str. **378, 67** (1996)
13. S. Jayanty, T.P. Radhakrishnan, Chem. Mater, **13,** 2460 (2001)
14. H.S. Mazumdar, S.R. Shah, Int. J. Res. in En. and Technol. 3(1), 2319 (2015)
15. P. Ert, B. Rohde, J. Cheminformatics, **4,** 12 (2012)
16. O. Dolomanov, Commission on Crystallographic Computing, IUCr Newsletter, **7**, 69 (2006)
17. A. Spek, IUCr Newsletter, **7**, 79 (2006)
18. A. Spek, IUCr Newsletter, **7**,78 (2006)
19. A.V. Dunaev, A.V. Larchenko, A.V. Buhanovskij, Scientific and Technical Statements of Saint-Petersburg State Polytechnic University (in Russian – Nauchno-tehnicheskie Vedomosti SPbGPU), **5**, 98 (2008)
20. K. Mamaev, A. Obkhodsky, A. Popov, IOP Conf. Series: Mater. Sci. and Eng. **112** (2016)
21. A.Yu. Kuznecov, A.B. Sobolev, A.S. Makarov, A.N. Velichko, Solid State Phys. (in Russian – Fizika Tverdogo Tela), **11**, 1950 (2005)
22. O. E. Gluhova, I. V. Kirillova, I. N. Salij, A. S. Kolesnikova, E. I. Kossovich, M. M. Slepchenkov, A. N. Savin, D. S. Shmygin, Math Modeling (in Russian – Matematicheskoe Modelirovanie), **9**, 106 (2012)
23. Yu. B. Malyhanov, R. N. Pravosudov, V. V. Meshkov, J. Structural Chem. (in Russian – Matematicheskoe Modelirovanie), **2**, 2017 (2000)
24. A.B. Gordienko, A.V. Kosobutckii, Solid State Phys. (in Russian – Fizika Tverdogo Tela), **58**, 3 (2016)

AUTHOR INDEX

EDP Sciences – Web of Conferences
17, Avenue du Hoggar
Parc d'Activité de Courtabouf
BP 112
F-91944 Les Ulis Cedex A
France

ISBN 978-1-5108-3564-1

CURRAN ASSOCIATES INC.
proceedings.com

International Conference on Big Data and its Applications (ICBDA 2016)

ITM Web of Conferences Volume 8 (2016)

Moscow, Russia
19 September 2016

edp sciences